Lighting

Lighting FOURTH EDITION

David C Pritchard BSc, CEng, MIEE, FCIBSE, MILE

Longman Scientific & Technical,
Longman Group UK Limited,
Longman House, Burnt Mill, Harlow,
Essex CM20 2JE, England
and Associated Companies throughout the world.

First published 1969
Second edition 1978
Third edition 1985
Second impression published by Longman Scientific & Technical 1987
Fourth edition 1992
Third impression 1994

British Library Cataloguing in Publication Data
Pritchard, D. C. David Christopher, *1928–*
Lighting. — 4th ed.
1. Buildings. Lighting
I. Title
729.28

ISBN 0-582-04655-6

Printed in Malaysia by PA

Contents

Acknowledgements

Figure 6.2 and Tables 6.1 to 6.8 are reproduced by courtesy of the Director, Building Research Establishment and by permission of the Controller, HMSO, Crown copyright. Extracts from BS 5489 are reproduced with the permission of BSI. Complete copies of the BS can be obtained by post from BSI Sales, Linford Wood, Milton Keynes, MK14 6LE: or by telephoning BSI Orderline, 0908 220022. Figures 1.6, 7.1, 7.7 and Table 6.9 are reproduced by permission of the Chartered Institution of Building Services Engineers, Delta House, 222 Balham High Road, London SW12 9BS. Tables 10.1 and 10.2 are reproduced by courtesy of Philips Lighting Ltd. Figure 4.5 is reproduced by courtesy of Sylvania Lighting Products Ltd. Figures 7.12, 7.13, 8.10 and 9.14 are reproduced by courtesy of Thorn Lighting Ltd.

The author would like to thank all those who have made helpful comments on the previous editions, in particular his colleague John F. Pickup.

1 The language of light

People's attitudes to lighting design are paradoxical. Those who have never considered the subject in any depth accept it as part of life and living, requiring little more thought than the strategic placing of some 60 watt 'bulbs' and the operation of a switch. And, to a point, they are right. After all, there are very few situations where there is insufficient light by which to see. The danger occurs when a person with this attitude has to make responsible lighting decisions on behalf of other people. The paradox is that as soon as a person becomes involved in lighting he enters a new science which demands his imagination and engineering ability. To be successful, he must be creative in his ability to reveal and display the visual scene. He must also have confidence in his calculations and an understanding of the lighting tools at his disposal. It is perhaps little wonder that electric lighting seldom compares favourably with daylight.

The first question could well be: 'What is light?' Strictly speaking, it is purely a human sensation in similar fashion to sound, taste, smell and warmth. Something is necessary to stimulate the senses, and in this case it is electromagnetic radiation falling on the retina of the eye. Light can therefore be considered as a combination of radiation and our response to it.

Electromagnetic radiation

The basic atomic structure is a positively charged nucleus around which orbits an array of negatively charged electrons, as shown in Fig. 1.1. If the atom receives energy by collision with another atom or by heat, it will release this energy via its electrons – hence *electro*magnetic. This radiation obeys various 'laws', travels in straight lines and consists of individual packets of energy called **photons**. The energy of a photon is

$$eV = h\nu \qquad \ldots [1.1]$$

where h is Planck's constant, ν is the frequency of the radiation, and eV is the energy in electron-volts, i.e. the work required to move an electron through a potential difference of one volt.

Example

A mercury electron emits photons whose energy level is 4.89 electron-volts. If Planck's constant is 6.62×10^{-34}J s calculate the wavelength of the emitted radiation.

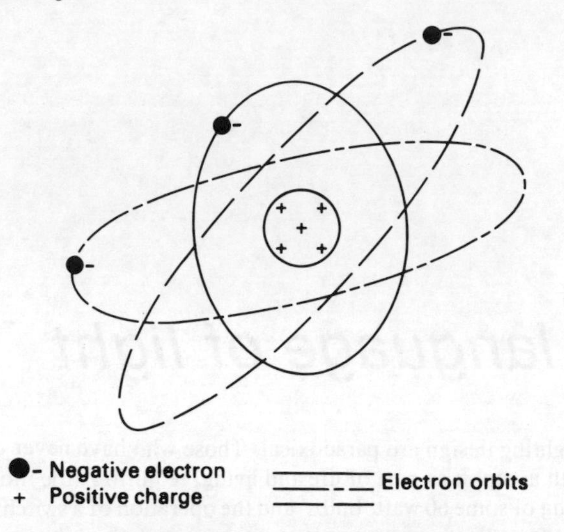

Fig. 1.1 Pattern of orbitting electrons in an atom

Solution

The frequency of the radiation in **hertz** (cycles per second), abbreviated Hz, can be found from the formula:

$$\nu = \frac{eV}{h}$$

When substituting values, the units must be correct. If h is in joule seconds, then eV must be in joules.

$$1 \text{ eV} = 1.603 \times 10^{-19} \text{ J}$$

$$\therefore \ \nu = \frac{4.89 \times (1.603 \times 10^{-19}}{6.62 \times 10^{-34}}$$

$$= 1.18 \times 10^{15} \text{ Hz}$$

To convert to wavelength, for all electromagnetic radiation

$$\nu = \frac{c}{\lambda} \qquad \ldots [1.2]$$

where c is the velocity of the photon (approx. 3×10^8 m/s)
λ is the wavelength in metres.

$$\text{Hence } \lambda = \frac{c}{\nu}$$

$$= \frac{3 \times 10^8}{1.18 \times 10^{15}} \text{ m}$$

$$= 254 \times 10^{-9} \text{ m}$$
$$= 254 \text{ nm}$$

This is the wavelength of the strong short wave ultra-violet radiation from mercury discharge.

Electromagnetic spectrum

Electrons are the source of radiation of varying frequencies and wavelengths and this radiation is used for many purposes other than light. It helps to decide whether to think in terms of wavelength or frequency.

As already seen in eqn [1.2]:

$$\text{Wavelength} \propto \frac{1}{\text{Frequency}}$$

In this book wavelength will be used to avoid confusion. The spectrum or range of wavelengths is shown in Fig. 1.2. The names of the various bands of the spectrum indicate, in general, how the energy is used. Light occupies a very narrow band of this spectrum (between 350 nm and 750 nm) and that band can be further subdivided into the range of colour sensations that most people experience.

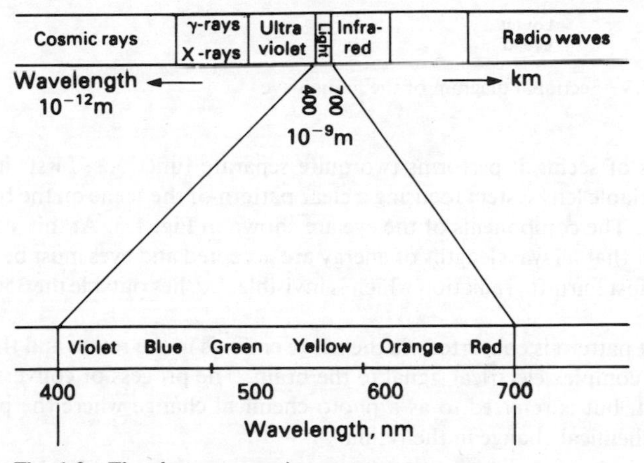

Fig. 1.2 The electromagnetic spectrum

One interesting aspect is the energy of a photon. This was given in eqn [1.1] and can be rewritten as

$$\text{Energy} \propto \frac{1}{\text{Wavelength}} \qquad \ldots [1.3]$$

Hence, as the wavelength gets shorter the energy of the photon increases. If this is related to the effect of radiation on people, the results can be quite devastating at wavelengths shorter than 300 nm. At longer wavelengths, in the millimetre and metre bands, the effect is negligible until it is transduced by a radio or television receiver.

Visual response

The eye is a magical part of the body. It conveys the bulk of our information and is the most precious of our five senses.

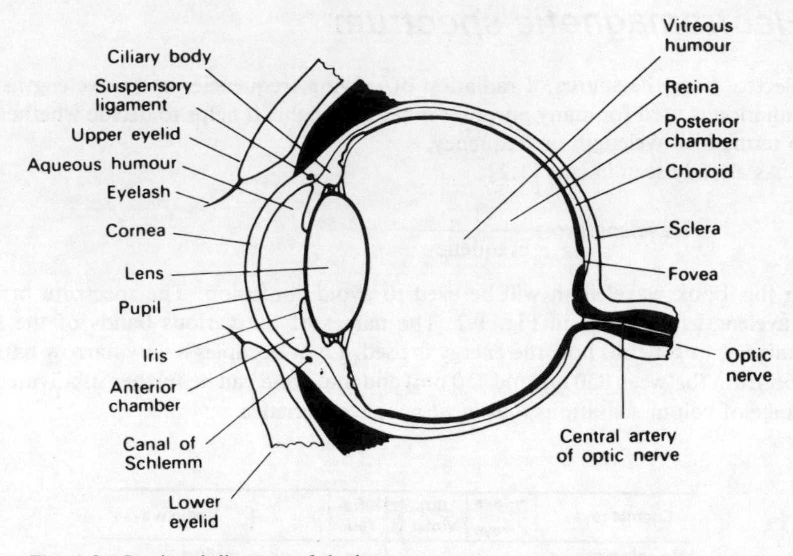

Fig. 1.3 Sectional diagram of the human eye

In the process of seeing it performs two quite separate functions. First, it acts as a sophisticated variable lens system focusing a clear pattern of the scene on the back of the eye – the retina. The components of the eye are shown in Fig. 1.3. At this stage it is as well to remember that all wavelengths of energy are accepted and eyes must be protected in some way against harmful radiation which is invisible, i.e. lies outside the 350–750 nm band.

Secondly, that pattern is converted via the nerve endings in the retina and through the optic nerve as a complex electrical signal to the brain. The process of conversion is not fully understood, but is referred to as a photo-chemical change where the photons of energy cause a chemical change in the retina.

Relative response of the eye

The nerve endings in the retina are of two types, cones (photopic vision) and rods (scotopic vision). The latter are highly sensitive and only effective at very low lighting levels, well below those to be discussed in this book, and will not be considered further.

The cones respond over a wave band from 400 to 730 nm and this response varies in both quantity and quality. Table 1.1 gives the relative response in terms of quantity, and the qualitive response in terms of colour has been shown in Fig. 1.2.

Table 1.1 shows (as one would expect) that there is a gradual increase to maximum response at 555 nm and then a gradual decrease. Plotting λ against V_λ will illustrate this and provide a useful reference curve.

Converting from radiation to light units

The radiation entering the eye is measured in watts. It then undergoes transformation because some wavelengths are more effective than others. The unit of light power is the **lumen** (abbreviated lm) and the general relation between the two units is:

Table 1.1 The luminous efficiency functions for photopic vision (V_λ)

λ (nm)	V_λ	λ (nm)	V_λ
400	0.000	550	0.994
410	0.001	560	0.995
420	0.004	570	0.952
430	0.011	580	0.870
440	0.023	590	0.757
450	0.038	600	0.631
460	0.060	610	0.503
470	0.090	620	0.381
480	0.139	630	0.265
490	0.208	640	0.175
500	0.323	650	0.107
510	0.503	660	0.061
520	0.710	670	0.032
530	0.862	680	0.017
540	0.954	690	0.008
		700	0.004
		710	0.002
		720	0.001
		730	0.000
		740	0.000

These values are from *Principles of light measurements, CIE Publication No. 18, 1970.*

$$\text{Light power} = K \int_{\lambda=\infty}^{\lambda=0} P_\lambda V_\lambda \text{ lm} \qquad \ldots [1.4]$$

where K is a factor normally accepted as 675
P_λ is the radiation power at wavelength λ
V_λ is the luminous efficiency of the eye at wavelength λ (see Table 1.1).

Example
If a light source emits 60 W at 589 nm, how many lumens does it emit?

Solution
At 589 nm, P_λ is 60 W (given in the problem) and V_λ is 0.757 (from Table 1.1). Therefore,

$$\text{Light power} = 675 \times 60 \times 0.757 \text{ lm} = 30\,700 \text{ lm}$$

This is a simple calculation when considering a single wavelength and requires more thought when the source is not a single wavelength (monochromatic) but a range of wavelengths (heterochromatic)

Example
Calculate approximately the lumens emitted from the following table of power distribution for a 200 W filament lamp.

Wave band (nm)	<400	400–450	450–500	500–550
Power (W)	2	1	1.5	2

Wave band (nm)	550–600	600–650	650–700	>700
Power (W)	2.5	3	3.5	170

Solution
This can be done in tabular form. The degree of accuracy will depend on the number of bands taken. In this case 50 nm bands are really too wide but an approximate answer is asked for.

Wave band (nm)	$P_\lambda(W)$	*Mid value* V_λ	$P_\lambda \times V_\lambda$
<400	2	0	0
400–450	1	0.008	0.008
450–500	1.5	0.110	0.160
500–550	2.0	0.780	1.560
550–600	2.5	0.910	2.270
600–650	3.0	0.320	0.960
650–700	3.5	0.020	0.070
>700	170	0	0
		Total =	5.01

$$\therefore \text{ Light power} = 675 \times 5.01 \text{ lm} = 3382 \text{ lm}$$

The efficiency of the lamp is expressed as **luminous efficacy** and is the ratio of lumens to lamp watts. In this case:

$$\text{Luminous efficacy} = \frac{3382}{200} = 16.9 \text{ lm/W}$$

It is worth noting that although a filament lamp is an efficient radiator, the large proportion of this radiation is beyond 700 nm in the infra-red band and hence useless. Filament lamps are basically inefficient.

Visual performance of the eye

A number of factors effect our ability to see and these can be subdivided into the human factors and the environmental factors. Human factors include:

(1) optical performance of the eye;
(2) colour perception;
(3) general condition of the eye.

Environmental factors include:

(1) relative brightness of the task and its surrounds;
(2) glare from the task or surrounds;
(3) movement in the task.

Optical performance of the eye

When the lens system is normal the eye should recognise detail of angular size 0.5 minutes

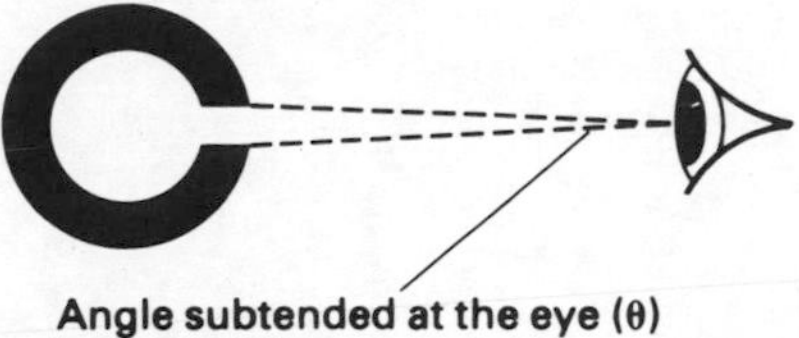

Fig. 1.4 The Landolt broken circle

of arc or less. 'Angular' size is the 'size' as seen and is both a function of size and viewing distance. This is illustrated in Fig. 1.4 using a standard text symbol called a 'Landolt broken circle'. An optician will test a person's sight by his ability to recognise letters on a standard test chart. A person who has 6 : 6 vision can recognise a 6 mm letter at a distance of 6 m; and 6 : 12 means he can only read at 6 m what a 6 : 6 can read at 12 m, i.e. his eyesight is substandard.

In experiments on vision and visual performance the process of recognition is normally described in terms of **visual acuity** where:

$$\text{Visual acuity} = \frac{1}{\text{Angular size (in minutes)}} \qquad \ldots [1.5]$$

Therefore, the starting point in understanding lighting is to appreciate that the eye must be functioning properly.

Example

A signwriter has to erect a nameboard with the letters BLC easily discernible from a distance of at least 1 km. Allowing for average acuity of 2 and adding a safety factor of 2, determine the minimum height of the letters if the gap in the letter C is one-fifth the height of the letter.

Solution

For a visual acuity of 2

$$2 = \frac{1}{\theta'} \qquad \ldots [*]$$

where θ is the angle subtended by the gap at the eye.

Let the height of the gap be x metres. If the angle θ is in radians, then

$$\theta = \frac{x}{\text{Distance}} = \frac{x}{1000}\ \text{rad}$$

but θ is in minutes and π radians is the same as 180 × 60 min. From [*] above, $\theta = \frac{1}{2}$ min; thus, converting minutes to radians,

$$\theta = \frac{1}{2} \times \frac{\pi}{180 \times 60}\ \text{rad}$$

$$= 0.000\,145\ \text{rad}$$

$$\therefore \quad \frac{x}{1000} = 0.000\,145\ \text{rad}$$

and $\qquad x = 0.145$ m

Increasing by the factor of 2
$x = 0.29$ m
$\therefore$ Letter height $= 5x$
$= 1.45$ m

Colour perception

There have been many theories on colour vision, the Trichromatic theory being the one most widely accepted today. In the chapter dealing with colour, the process of additive mixing of primary colours is discussed and it is shown that this can produce a complete spectrum of colours. Applying this to vision, the theory considers the cones to be grouped into three types, each type sensitive to different wavelengths of energy. The interaction of these three groups is then responsible for the stimulus which is interpreted by the brain as 'colour'. We may not all see colours in the same way although we will all describe energy at, for example, wavelength 650 nm as creating a 'red' sensation. This particular consideration may help to explain why people can have such varying tastes in colour.

Despite the fact that, under normal day vision, colour is always present, the units and design techniques in lighting calculations are not able to take colour fully into account. This is not an implication that colour is unimportant.

An understanding of colour is essential when discussing visual performance, for not only are some colours more visible than others, but the contrast between colours is important. Blue contrasts strongly with yellow as these colours are 'complementary', but not as strongly with green as these colours are close in the spectrum.

General condition of the eye

The eye deteriorates with age. Anyone born with normal eyesight can expect to be wearing glasses by the time they reach 40. This is mainly due to weakening of the lens muscles and a difficulty in being able to focus at short distances. This can be partially corrected by wearing glasses. The condition is known as **presbyopia**. More serious problems include cataract, which is the clouding over of the lens. This impairs clear vision and scatters the light entering the eye making the sufferer far more sensitive to glare.

It is difficult to take these kinds of defects into account in general lighting schemes, but where it is known that the average age will be high – e.g. an old people's home – then extra consideration must be given to the quantity of light and avoidance of glare.

Relative brightness-contrast sensitivity

An object is seen because of the difference in brightness between the object and its immediate background. For example, a white letter would be very clearly seen against a black background but would be virtually invisible against a white background. This is an obvious statement, but the visual process is not so obvious as to suggest that we always strive for maximum contrast. People who wear tinted sun-glasses do not think so, as they deliberately reduce the brightness of the backgrouind.

There are two main factors to consider:

1. The contrast between the task and its immediate surrounds, e.g. the letters and page of this book.
2. The brightness of the general background, e.g. the table on which this book lies or what is seen around the book when reading.

When discussing mathematical relationships, objective terms are normally used. In this instance **brightness** is subjective and describes the appearance, **luminance** is objective and is the physical measurement of brightness. The unit of luminance is the candela per square metre (abbreviated cd/m²)

$$\text{Contrast} = \frac{L_T - L_B}{L_B} \qquad \ldots [1.6]$$

where L_T is the luminance of the task
L_B is the luminance of the immediate background.

Example
Dark wool is viewed against a dark background. The illumination is the same on both wool and background. If the reflectance of the background is 3 per cent and that of the wool is 5 per cent, what is the contrast?

Solution
If the illuminance is constant, then the luminance is proportional to the reflectance, and from eqn [1.6]

$$\text{Contrast} = \frac{0.5 - 0.3}{0.3}$$

$$= 0.66$$

Contrast sensitivity is more difficult to explain. It refers to the ability to recognise contrast in the task and is basic to work on contrast rendition factors (CRF). Considering the threshold condition – i.e. when the contrast is just visible – the difference in luminance can be expressed as ΔL. This again can be expressed as a ratio to the average luminance. Figure 1.5 shows an experimental relation between this threshold contrast and the general luminance of the field of view. It indicates an increase in sensitivity as the general luminance increases. What it does not show is that further increases in luminance will tend to decrease the contrast sensitivity due to the effects of glare.

Example
The general lighting level is increased to raise the general luminance level from 10 to 100 cd/m². What is the change in contrast sensitivity expressed in terms of contrast?

Solution
Using eqn [1.6] and considering the threshold condition, it can be deduced from Fig. 1.5 that

(1) at 10 cd/m² the contrast is approximately 0.9 per cent; and
(2) at 100 cd/m² the contrast is approximately 1.1 per cent.

This indicates the minimum contrast that can be seen under specified lighting conditions.

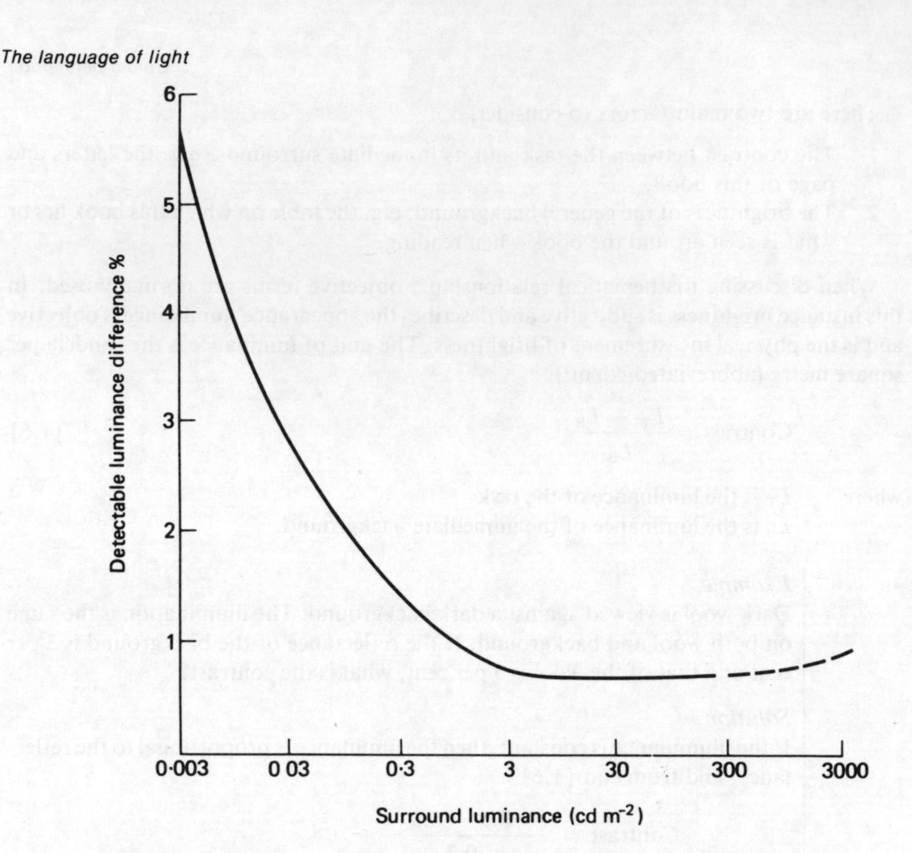

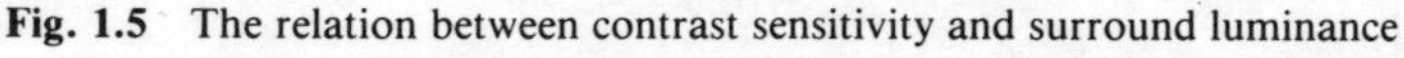
Fig. 1.5 The relation between contrast sensitivity and surround luminance

Applying eqn [1.6] to condition (1):

$$L_B = 10\ \text{cd/m}^2$$

the minimum perceptible contrast difference $\Delta(L_T - L_B)$ is given by

$$\frac{\Delta(L_T - L_B)}{L_B} = \frac{0.9}{100}$$

$$\therefore \quad \Delta(L_T - L_B) = 10 \times \frac{0.9}{100}$$

$$= 0.09\ \text{cd/m}^2$$

Applying eqn [1.6] to condition (2):

$$L_B = 100\ \text{cd/m}^2$$

$$\frac{\Delta(L_T - L_B)}{L_B} = \frac{1.1}{100}$$

$$\therefore \quad \Delta(L_T - L_B) = \frac{100 \times 1.1}{100}$$

$$= 1.1\ \text{cd/m}^2$$

This shows that, as the lighting level is increased, the minimum perceptible luminance difference is increased. This is of little significance if the lighting is increased evenly. If the background luminance is increased, but not the task – e.g. by working on a white surface rather than a grey one – then this increase from 0.9 to 1.1 cd/m² represents a loss in visual performance. It is these aspects of visual performance that concern research workers such as H. C. Weston, H. R. Blackwell, P. Boyce and others. Specifying good lighting conditions is more than specifying the quantity of light. It involves the 'revealing power' in terms of the direction of light, surface properties of the task, and general lighting of the environment.

Glare from the task and its surrounds

Glare is a development of contrast. Normally the eye adapts to whatever it is viewing, but if the task or background are too bright or the contrast is too great, vision suffers either by the situation becoming visually uncomfortable (**discomfort glare**) or by the task becoming difficult to see (**disability glare**), or both. Examples of glare unfortunately surround us, and can be far more disturbing than lack of illumination.

Common forms of discomfort glare are interior and exterior electric lighting installations. Chapter 7 explains how interior lighting schemes can be designed to keep this type of glare under control.

Disability glare can come direct from a light source or window, or by reflection off a glossy surface (**velling glare**). The modern office with its array of visual display units is a common situation where this problem must be considered. CRF, referred to above, is one aspect of design where veiling glare is recognised and accounted for in the design process.

Movement in the task

Seeing is not a static process. The eye continually scans the field of view. We blink approximately 15 times per minute and the retinal response automatically adapts to the general lighting level. The lens shape changes to alter the focus and the iris expands and contracts to control the amount of light entering the eye.

The source of light often varies in quantity and quality, whether daylight or electric light. If electric, it will probably be flashing 100 times per second. So even in what appears to be a passive situation, such as sitting at a desk reading, there is a considerable element of movement and change.

Normal assessment of lighting requirements assumes that the task is stationary and there is sufficient time to focus on it. When the task is moving, such as work on a conveyor belt assembly, then extra lighting should be provided.

Visual task requirements

The two main aims of good lighting are to:

(1) effectively reveal the task;
(2) appropriately reveal the general surround.

To achieve (1) involves consideration of the factors already discussed, and Fig. 1.6 (taken from the CIBSE 1984 Lighting Code) shows how these factors can be related. This

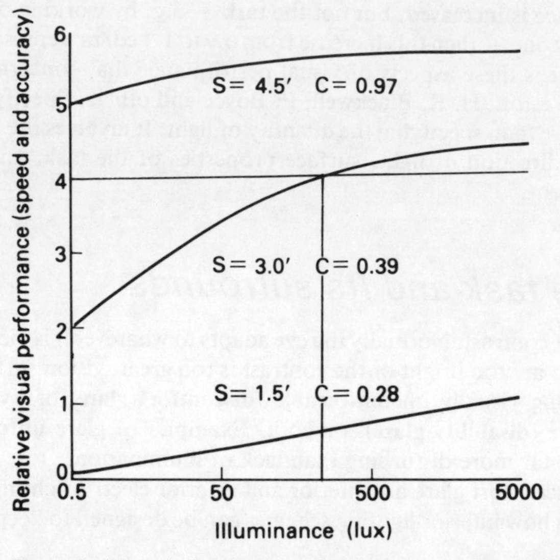

Fig. 1.6 The relation between visual performance and illuminance for task sizes and contrast

particular diagram relates visual performance to illuminance levels for three specific contrasts (C) and angular sizes (S). The visual performance scale is relative and, if the task is very small and the contrast poor, there always seems to be room for improvement. If the task is large and the contrast good, then even moonlight could be sufficient to see by. This type of information is the basis for the recommended lighting levels quoted in codes and standards.

Example

A task comprises reading detail 0.5 mm wide at 0.5 m viewing distance. The reflectance of the task is 0.3 and the immediate background 0.2. What lighting level should ensure a relative visual performance of 4?

Solution

The contrast can be found from the reflectances

$$C = \frac{0.3 - 0.2}{0.2}$$

$$= 0.5$$

Figure 1.6 uses angular size in minutes of arc. In this case:

$$\text{Angular size} = \frac{0.5}{500}$$

$$= 0.001 \text{ rad}$$

$$= 0.001 \times \frac{180 \times 60}{\pi} \text{ min}$$

$$= 3.43 \text{ min}$$

This particular curve is not on Fig. 1.6 but can be deduced as shown by the dotted line. This will give an illuminance around 300 lx.

Visual preference

Man normally has the right to express his own wish or preference as to how his living conditions should be. This is partly based on what he might expect to have in relation to those around him and partly on the minimum that he knows he needs. Motorists may prefer to have a good performance, top of the range family car because this would compare favourably with those around them. They know that a basic 3-door hatchback would meet their needs. They would, however, consider a 3-litre luxury saloon car well beyond their needs.

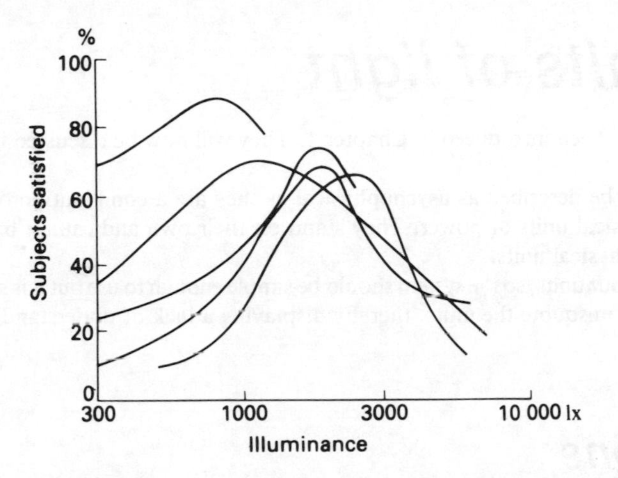

Fig. 1.7 Curves from various research reports on office workers' visual preference of illuminance at work (based on CIE Publication No. 29, 1975)

The same range of choice can be applied to lighting. Figure 1.7 gives a summary of the results of different researches when offering people a free choice of office lighting. Too many other factors are involved to draw conclusions beyond the fact that people prefer generous lighting standards, but this may not be sufficient justification to grant their wishes.

2 Units of light

Various units have been introduced in Chapter 1. They will now be discussed in greater detail.

Light units can be described as **psychophysical** as they are a combination of human response and physical units of power. They stand on their own and cannot be directly related to other physical units.

There are only four units, so the system should be simple enough to use but it is surprising how many people misquote the units, thereby displaying a lack of understanding.

Definitions

The following definitions are based on BS 4727 Part 4: *Glossary of terms particular to lighting and colour*. The units are in the International System as defined in British Standard 3763.

Luminous flux (symbol ϕ; unit, lumen)

The light *emitted by a source*, or received by a surface. The quantity is derived from radiant flux (power in watts) by evaluating the radiation in accordance with the relative luminous efficiency of the 'standard' eye (V_λ).

Lumen (symbol, lm)

The SI unit of luminous flux.

Luminous intensity (symbol, I; unit, candela)

The quantity which describes the power of a source or illuminated surface to emit light in a given direction.

Note: 1. One may also speak of the mean intensity in a group of directions.
2. Only for a source of *uniform* intensity is it not required to specify direction.

Candela (symbol, cd)

The SI unit of luminous intensity, equal to one lumen per steradian.

Illuminance (symbol, E; unit, lux)

The luminous flux density at a point on a surface, i.e. the luminous flux incident per unit area.

Lux (symbol, lx)

The unit of illuminance; one lumen per square metre.

Luminance (symbol, L; unit, candela per square metre)

The intensity of the light emitted in a given direction per projected area of a luminous or reflecting surface.

Note: Only for a source of *uniform* luminance is it not required to specify direction.

Candela per square metre (symbol, cd/m^2)

The unit of luminance; one candela per square metre of a projected area.

Discussion of units

Light rays travel in a straight line. As the light rays will be diverging, the amount of light incident per unit area will decrease as the receiving surface moves away from the light source.

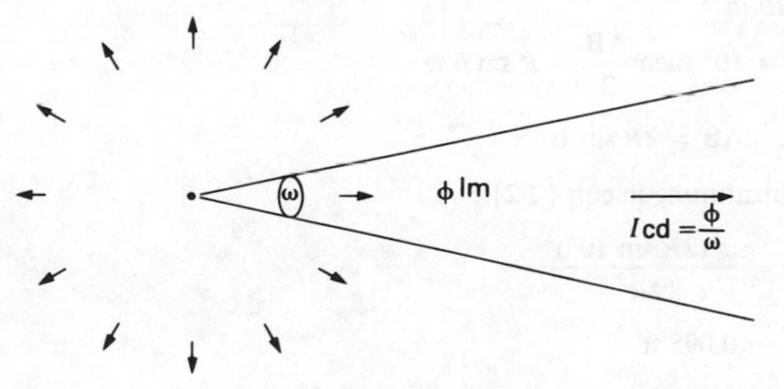

Fig. 2.1 Relation between flux and intensity for small source

Figure 2.1 shows a light source emitting equally in all directions. It also shows that the relation between the flux and the uniform intensity in a solid angle is given by

$$\phi = I \times \omega \qquad \ldots [2.1]$$

Light is radiated in three dimensions even though, for convenience, data are recorded in terms of two dimensions. The surface of this paper is two-dimensional, it has breadth and width. A distant point and the edges of the paper form a pyramid, a three-dimensional figure containing a given solid angle.

In a similar fashion a plane angle can be drawn on a flat surface and can be referred to as a *planar angle*, measured in radians or degrees. A solid angle can only be drawn to suggest a solid and is measured in *steradians* (st).

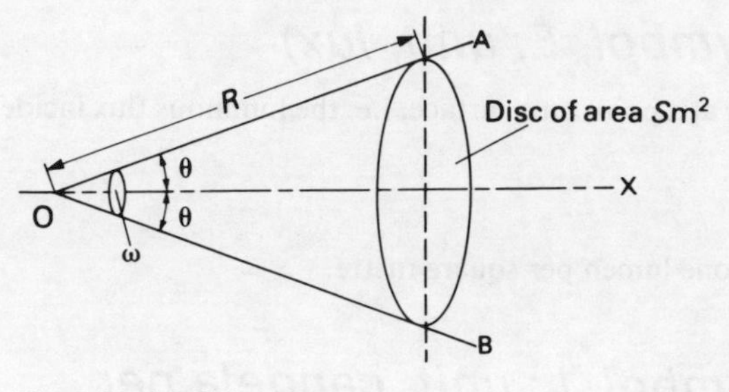

Fig. 2.2 A solid angle

A solid angle is illustrated in Fig. 2.2.

$\widehat{AOB}$ is a plane angle 2θ. Spin $\widehat{AOB}$ around OX and a cone is formed.

$$\widehat{AOB}\text{ (in radians)} = \frac{AB}{R}\text{ (provided }\widehat{AOB}\text{ is small)}$$

$$\omega\text{ (solid angle)} = \frac{\pi\,(AB)^2}{4R^2}\text{ approx.} \qquad \ldots [2.2]$$

If θ is 90° or $\pi/2$ rad, ω is 2π st.

If θ is 180° or π rad, ω is 4π st.

Example

If, in Fig. 2.2, θ is 10°, what is the value of ω?

Solution

If $\theta = 10°$ then $\frac{AB}{2} = R \sin\theta$

$\therefore \quad AB = 2R \sin 10°$

Substituting in eqn [2.2]

$$\omega = \frac{\pi\,(2R \sin 10°)^2}{4R^2}$$

$$= 0.095 \text{ st}$$

Luminous intensity

The concept and definition of 'luminous intensity' may be difficult to grasp. Figure 2.1 shows luminous flux as being emitted within a solid angle, which suggests a cone of substantial size. On the other hand, the definition of luminous intensity considers a cone of such small dimensions that it becomes a single line, i.e. a specific single direction.

Flux in a solid angle is the product of the mean luminous intensity and a solid angle. If the solid angle becomes so small that it is associated with a single direction – i.e. if the solid angle approaches zero – it is difficult to conceive that there is any flux at all! It is more helpful if we do not think of the solid angle as being infinitely small, but as being just small enough that light may be considered to be emitted evenly within it. Luminous intensity is the unit used to define the 'strength' of light in specific directions, and from that unit we can derive the other three:

Luminous flux. This is derived from eqn [2.1] and is dimensionally the same as intensity.

Illuminance. This is luminous flux falling on a unit area of surface and, hence, is related to intensity.

Luminance. Equation [2.1] does not define the size of the source emitting lumens. If it is a flat area S m^2, then

$$L = \frac{I}{S} \text{cd/m}^2 \qquad \ldots [2.3]$$

It is important to appreciate that S is the area *as seen* from the direction specified, called the **apparent** or **projected** area. Figure 2.3 shows some examples of apparent areas.

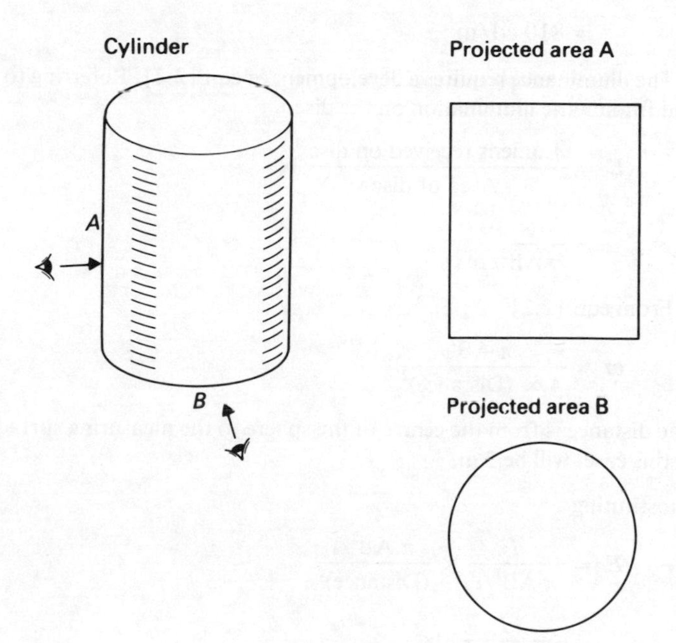

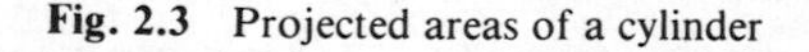
Fig. 2.3 Projected areas of a cylinder

Example
A sphere of diameter 0.5 m emits 2000 lm uniformly in all directions. Calculate the average intensity, luminance, and illuminance on a surface 3 m from its centre.

Solution
The average intensity is derived from eqn [2.1]. In this case the total solid angle 4π is taken since the emission of light is the same in all directions:

$$\phi = I \times \omega \text{ lm}$$

$$\therefore \quad I = \frac{\phi}{\omega}$$

$$= \frac{2000}{4\pi}\ \text{cd}$$

$$= 159\ \text{cd}$$

The luminance is derived from eqn [2.3]:

$$L = \frac{I}{S}\ \text{cd/m}^2$$

$$= \frac{159}{\text{Apparent area of sphere}}\ \text{cd/m}^2$$

$$= \frac{159}{\pi \times 0.25^2}\ \text{cd/m}^2$$

$$= 810\ \text{cd/m}^2$$

The illuminance requires a development of eqn [2.2]. Referring to Fig. 2.2, and finding the illumination on the disc,

$$E = \frac{\text{Lumens received on disc}}{\text{Area of disc}}\ \text{lx}$$

$$= \frac{I \times \omega}{\pi\ \text{AB}^2/4}$$

From eqn [2.2]

$$\omega = \frac{\pi\ \text{AB}^2}{4 \times (\text{Distance})^2}$$

The distance is from the centre of the sphere to the measuring surface which, in this case, will be 3 m.

Substituting

$$E = \frac{I}{\pi\ \text{AB}^2/4} \times \frac{\pi\ \text{AB}^2/4}{(\text{Distance})^2}$$

$$= \frac{I}{(\text{Distance})^2}\ \text{lx} \qquad \ldots [2.4]$$

(this equation is known as the Inverse Square Law of Illumination)

$$= \frac{159}{9}\ \text{lx}$$

$$= 17.7\ \text{lx}$$

Luminous flux

The lumen is used to define light output of lamps, e.g. a 60 watt GLS lamp emits 660 lm. It is also used to quote the luminous efficacy of lamps (see Chapter 1). Typical luminous efficacies are given in Table 2.1.

Table 2.1 Typical luminous efficacies

Light source	*Average Efficacy (lm/W)*
Low-pressure sodium lamp	180
High-pressure sodium lamp	105
Sunlight	80
White fluorescent tube	85
Tungsten – halogen lamp	22
GLS filament lamp	15

If the lumens in a specific zone are required – e.g. the upward lumens from a luminaire – then the calculation involves further knowledge of intensity. This will be considered in the next section.

Luminous intensity

Unless referring to mean or average intensity, this unit is always related to direction. Direction must normally be specified in terms of three axes. This raises problems when using flat diagrams as these can only show two of the axes. Plots of intensity distribution are shown on a flat diagram (polar co-ordinate graph) and this will refer to the pattern of intensity distribution in a specific plane.

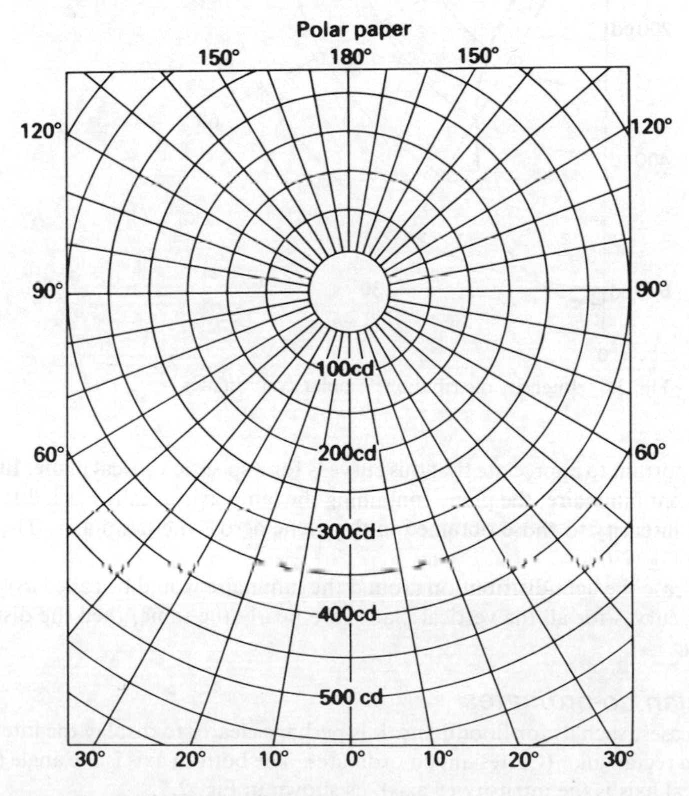

Fig. 2.4 Polar intensity diagram

Intensity distribution diagrams (polar curves)

The two variable dimensions are intensity (candelas) and direction (degrees). The simplest diagram for displaying this information is the polar curve (see Fig. 2.4), which is used to show the intensity distribution, usually in a vertical plane. The 'spokes' represent the various angles from the downward vertical and the circles the magnitudes of intensity. Figure 2.5 shows a typical curve for a half-plane, i.e. where the intensity distribution is common to all planes. It is only necessary to produce a half-plane polar curve. The point of origin of the angles is the photometric centre of the light source.

Example

The following table gives a list of intensities at different vertical angles. Plot the polar curve.

Angle from downward vertical (°)	0	15	30	45	60	75	90
Intensity (cd)	620	570	400	200	50	10	0

Solution

This is shown on Fig. 2.5.

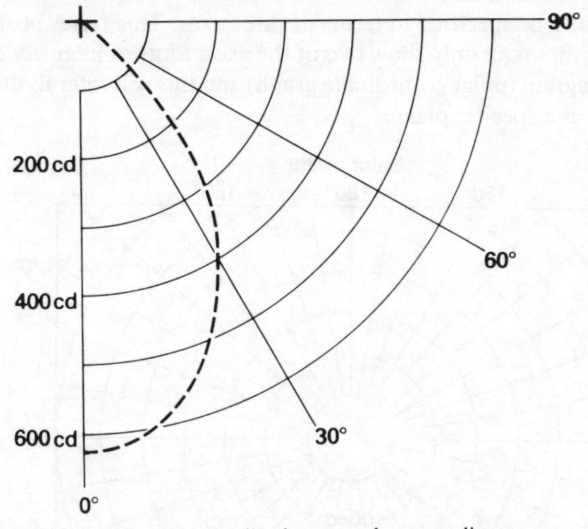

Fig. 2.5 Intensity distribution – polar co-ordinates

It is important to appreciate that this curve is for a specific vertical plane. In the case of a fluorescent luminaire, the plane containing the lamp axis usually has a different set of values of intensity to those obtained in the plane across the lamp axis. These axes are shown in Fig. 2.6.

In this case the light distribution around the luminaire would be called *asymmetric*. If the polar curves for all the vertical planes are nearly the same, then the distribution is *symmetric*.

Cartesian co-ordinates

In some cases, such as for floodlights, it is perhaps clearer to display the intensity distribution on rectangular (Cartesian) co-ordinates. The bottom axis is the angle (x axis) and the vertical axis is the intensity (y axis), as shown in Fig. 2.7.

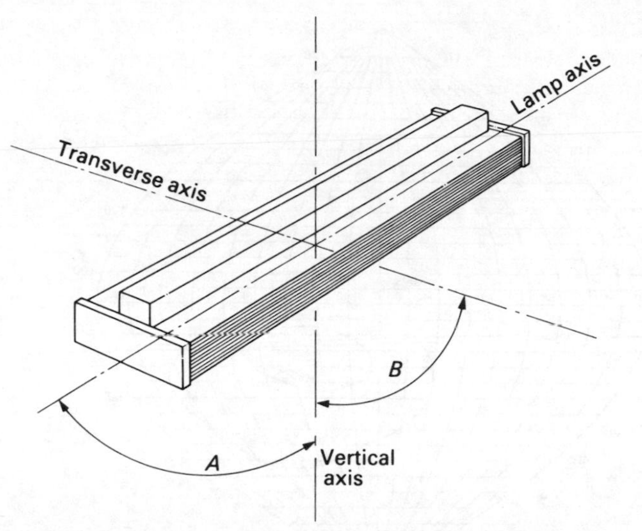

Fig. 2.6 Major vertical planes (courtesy Philips Lighting Ltd.)

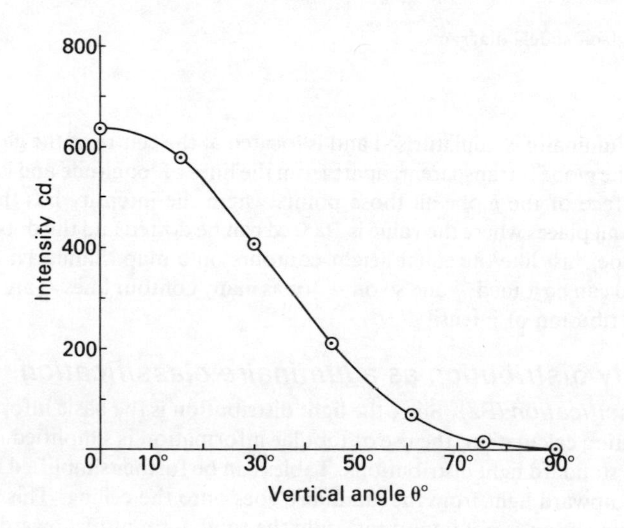

Fig. 2.7 Intensity distribution – Cartesian co-ordinates

Iso-candela diagram

The intensity distributions of some luminaires (e.g. road lighting lanterns) are so complex that a large number of plane polar curves would be required to record such a distribution.

Complex distributions are displayed on a web called an iso-candela diagram. With such a diagram, it is possible to display the intensities in all directions in a hemisphere.

Figure 2.8 shows such a web, which is similar to the web used by geographers to display the height contours of the earth's surface. Visualise the diagram as not flat, but hemispherical and curving upwards from the paper. The curves drawn upon it, roughly vertically and horizontally, correspond to the lines of longitude and latitude on a globe.

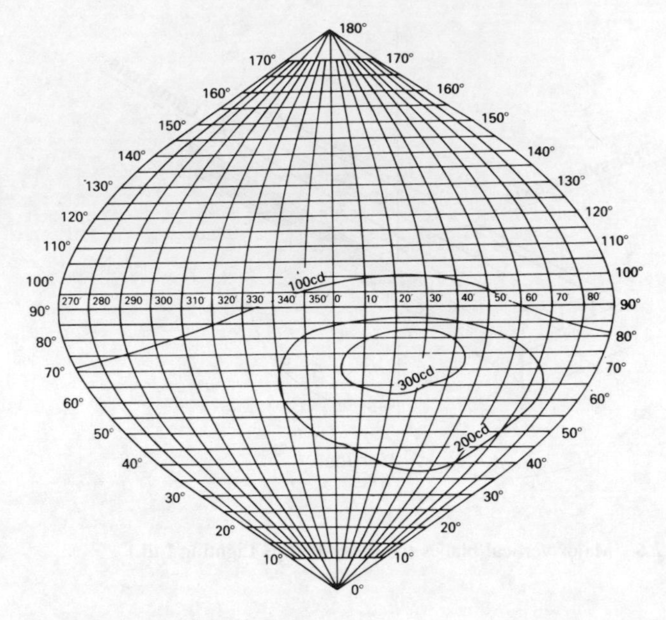

Fig. 2.8 Iso-candela diagram

Imagine that the luminaire is miniaturised and is located at the centre of the globe and that the surface of the globe is transparent, apart from the lines of longitude and latitude.

Mark on the surface of the globe all those points where the intensity has the same value; for example, all places where the value is 1000 cd can be dotted and the dots joined to give a contour line, just like the equal height contours on a map. Similarly, all dots representing 2000 cd can be joined – and so on – for as many contour lines as are needed to illustrate the distribution of intensity.

Use of intensity distribution as a luminaire classification

British Zonal classification (BZ). Since the light distribution is the basic information required in any lighting calculation, the use of tabular information is simplified if it can be based on a set of standard light distributions. Tables can be further simplified if it can be assumed that all upward light from the luminaire goes onto the ceiling. This implies that the upward distribution is not important, only the total amount of upward light.

Different countries have different classification techniques. The British Zonal system of the CIBSE classifies ten standard forms of downward light distribution which are:

BZ 1	$I_\theta \propto \cos^4 \theta$
BZ 2	$I_\theta \propto \cos^3 \theta$
BZ 3	$I_\theta \propto \cos^2 \theta$
BZ 4	$I_\theta \propto \cos^{1.5} \theta$
BZ 5	$I_\theta \propto \cos \theta$
BZ 6	$I_\theta \propto (1 + 2 \cos \theta)$
BZ 7	$I_\theta \propto (2 + \cos \theta)$
BZ 8	I_θ is constant
BZ 9	$I_\theta \propto (1 + \sin \theta)$
BZ 10	$I_\theta \propto \sin \theta$

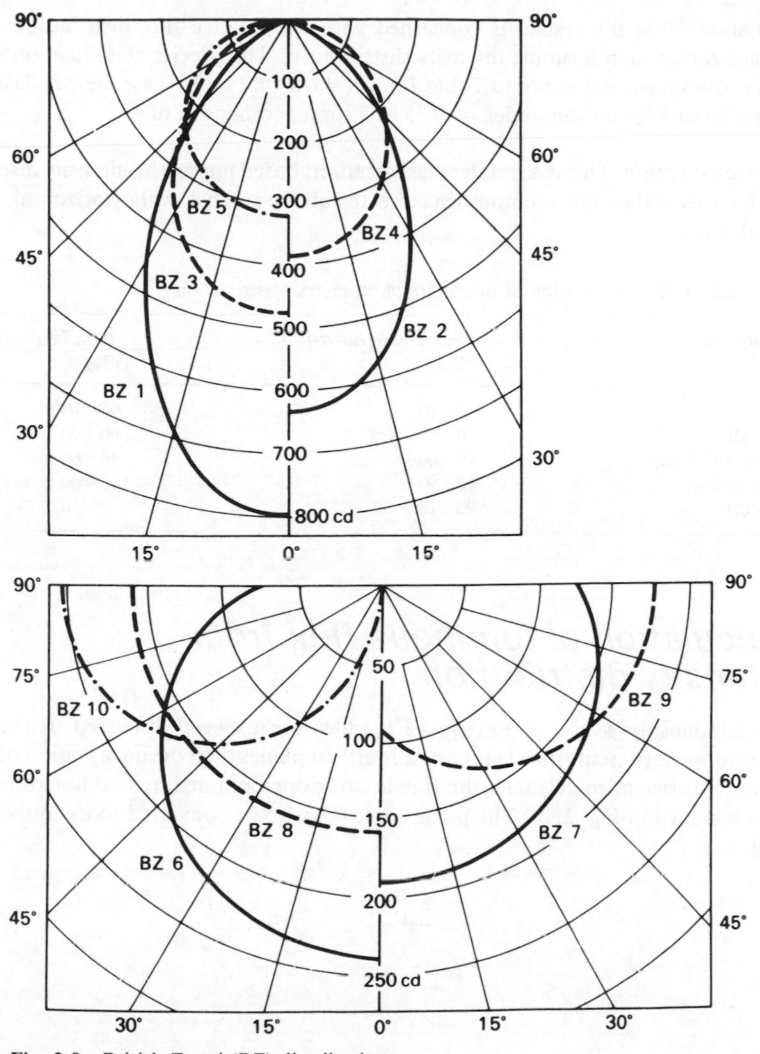

Fig. 2.9 British Zonal (BZ) distributions

I_0 is the intensity (in candelas) at 0°, directly below the luminaire; and I_θ is the average intensity at angle θ to the downward vertical. These curves are illustrated in Fig. 2.9.

Any light distribution can thus be specified in terms of its nearest BZ number. Sometimes it is difficult to decide which is the nearest number. The exact specification is beyond the scope of this book, but the relationship between light distribution and different room sizes has to be considered and the calculations are available in CIBSE Technical Memorandum No. 5.

The BZ system is defined as

> 'A system for classifying luminaires as described in CIBSE Technical Memorandum No. 5. The BZ class number (e.g. BZ 5) denotes the classification of a luminaire in terms of the flux from a conventional installation directly incident on the working plane, relative to the total flux emitted below the horizontal (the direct ratio).'

This shows that the system is concerned with the relative flux distribution on room surface rather than a simple intensity distribution. The precise classification may vary with room shape. Reference to Table 7.3 will show that in this case the BZ classification ranges from 3 for a room index of 0.75 to 5 for a room index of 5.

CIE classification. This is a simpler classification, based not on the intensity distribution, but in terms of the total luminous flux directed above and below the horizontal, as shown in Table 2.2.

Table 2.2 CIE classification of luminaire photometric distribution

Luminaire	*Flux above horizontal (%)*	*Flux below horizontal (%)*
Direct	0–10	90–100
Semi-direct	10–40	60–90
General diffuse	40–60	40–60
Semi-indirect	60–90	10–40
Indirect	90–100	0–10

Calculation of luminous flux from intensity distribution

The relationship $\phi = I \times \omega$ (eqn [2.1]) has already been discussed. If the intensity distribution represents the average of all vertical planes then the information of the polar curve is sufficient to calculate the flux in any zone bounded by two horizontal planes. This is shown in Fig. 2.10. The plane angle θ has to be converted to its equivalent solid angle ω.

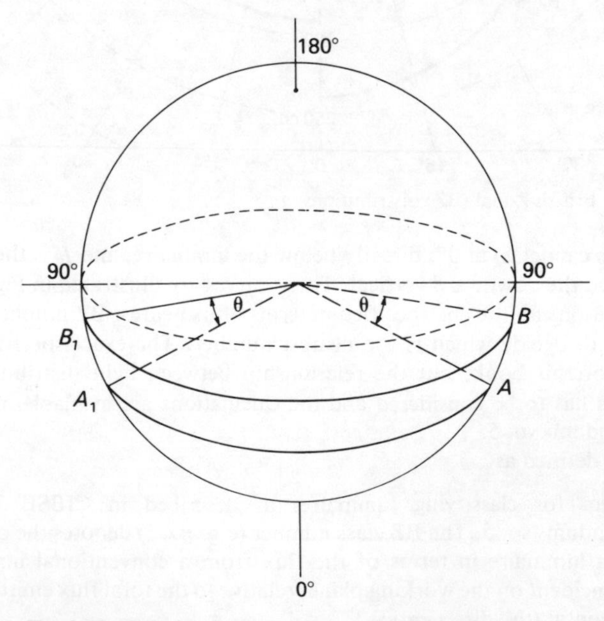

Fig. 2.10 Relation between plane and solid angle (zone factors)

The normal transformation is given by

$$\omega = 2\pi(\cos\theta_1 - \cos\theta_2) \qquad \ldots [2.5]$$

where θ_1 and θ_2 are the angles, as shown on Fig. 2.10.

This expression is referred to as the **zone factor** (ZF) and Table 2.3 shows the zone factors if the plane angle zones are at 10° intervals. The values from 90° to 180° will be the mirror image of those from 90° to 0°, i.e. ZF(80°–90°) = 1.091 = ZF(100°–90°).

Table 2.3 Zone factors

Vertical angle (θ)°	*Zone factor (ω)*	*Vertical angle (θ)°*	*Zone factor (ω)*
0–10	0.095	50–60	0.897
10–20	0.283	60–70	0.993
20–30	0.463	70–80	1.058
30–40	0.628	80–90	1.091
40–50	0.774		

Having now established a relationship between plane angle (θ) and solid angle (ω) for 10° zones, the mid-intensity values are determined for each zone, and hence the luminous flux in that zone, from:

$$\text{Luminous flux in zone } \theta_1 \text{ to } \theta_2 = \frac{I_{\theta 1} + I_{\theta 2}}{2} \times 2\pi(\cos\theta_1 - \cos\theta_2) \qquad \ldots [2.6]$$

Using zone factors it is possible to calculate the efficiency of luminaires quoted in terms of light output ratio.

$$\text{Light output ratio (LOR)} = \frac{\text{Total lumen output of luminaire}}{\text{Total lamp lumen output}}$$

$$\text{Downward light output ratio (DLOR)} = \frac{\text{Downward lumen output of luminaire}}{\text{Total lamp lumen output}}$$

$$\text{Upward light output ratio (ULOR)} = \frac{\text{Upward lumen output of luminaire}}{\text{Total lamp lumen output}}$$

$$\text{LOR} = \text{DLOR} + \text{ULOR}$$

Example

A luminaire has an intensity distribution of $100 \cos\theta$ (this is BZ 5) and emits no light above 90°. If the lamp emits 500 lm, what is the LOR of the luminaire?

Solution

This is best dealt with by constructing a table (see Table 2.4).

$$\text{Total light output} = \text{Sum of column D}$$
$$= 315\,\text{lm}$$

$$\text{Light output ratio (LOR)} = \frac{315}{500}$$
$$= 0.63$$

In this case, DLOR is also 0.63.

As there is no light above 90°, ULOR is 0.

Table 2.4

A *Zone* $(\theta_1 - \theta_2)°$	*B* *Average intensity* $I_0 \cos\left[\frac{\theta_1 + \theta_2}{2}\right]$ *cd*	*C* *Zone factor* $(2\pi(\cos\theta_1 - \cos\theta_2))$ *sr*	*D* *Flux* $B \times C$ *lm*
0–10	99.6	0.095	9.5
10–20	96.6	0.283	24.4
20–30	90.6	0.463	42.0
30–40	81.9	0.628	51.4
40–50	70.7	0.774	54.8
50–60	57.4	0.897	51.0
60–70	42.3	0.993	42.0
70–80	25.9	1.058	27.4
80–90	8.7	1.091	9.5
90–180	0	6.28	0

Illumination

The deducing of illuminance is the most common form of calculation that a lighting engineer has to perform. It is the most precise part of lighting specification and is the only calculation that can be checked, with any reasonable degree of accuracy, in a completed installation. There are two basic situations:

(1) the direct illuminance at a point on a specified plane (E_D);
(2) the average illuminance on a room surface due both to direct flux and that received by reflection off other surfaces (E_{av}).

Situation (1) will be considered in this chapter and (2) in Chapter 7.

Calculation of direct illuminance

The point source. This is an important concept. In theory a point source is infinitely small, but in practice it is a light source small enough to consider that all the light comes from a clearly defined point in space. This considerably simplifies the calculations and it is reasonable to assume that a light source is a point if the measuring distance is at least five times the maximum luminous width of the source. For instance, for a 1.5 m long fluorescent tube, the measuring distance would have to be at least 7.5 m (which is a far greater distance than normally experienced in a situation where illuminance calculations are needed).

The inverse square law. Referring to Fig. 2.11, the illuminance at a point A on a horizontal plane directly below the source is:

$$E = \frac{I_0}{H^2} \text{ lx} \qquad \text{... [2.7]}$$

where I_0 is the intensity towards point A (cd)
H is the height of the source above the plane (m)

Now considering point B, three things have changed:

1. The measuring distance is now PB.
2. The intensity is now I_θ.

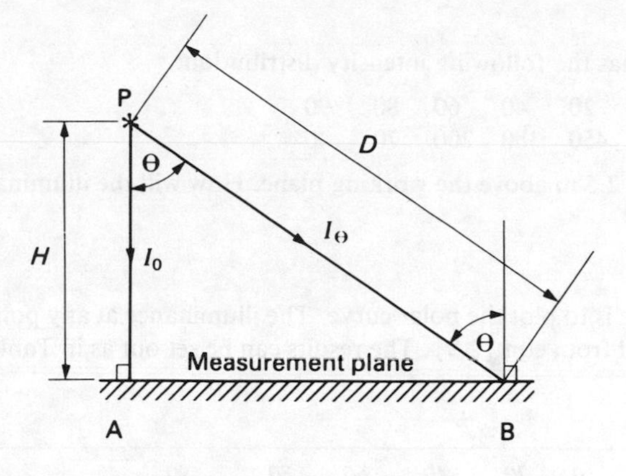

Fig. 2.11 Illuminance due to a small light source

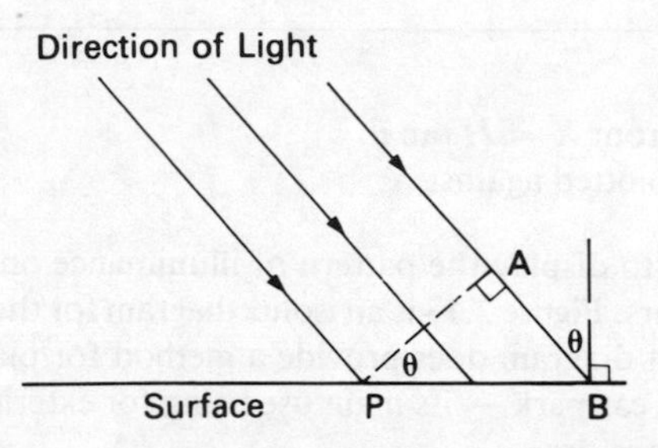

Fig. 2.12 Tilting the plane of illuminance – cosine effect

3. The surface is no longer at right angles or normal to the direction of the incident light; it is effectively tilted away from the normal by the angle θ and Fig. 2.12 indicates that the effective area has been increased by the ratio PB to PA, where

$$\cos\theta = \frac{\mathrm{PA}}{\mathrm{PB}}$$

Hence, the illuminance falls by the factor of $\cos\theta$, and the illuminance at B is

$$E = \frac{I_\theta}{\mathrm{PB}^2}\cos\theta\,\mathrm{lx} \qquad \ldots [2.8]$$

This can be simplified to two variables:

$$\cos\theta = \frac{H}{\mathrm{PB}}$$

$$\therefore \qquad E = \frac{I_\theta}{H^2}\cos^3\theta\,\mathrm{lx} \qquad \ldots [2.9]$$

This is sometimes called the $\cos^3$ 'law' of illumination.

Example

A luminaire has the following intensity distribution:

$\theta°$	0	20	40	60	80	90
I_θ(cd)	420	450	380	200	70	0

It is mounted 2.5 m above the working plane. How will the illuminance vary on this plane?

Solution

The first stage is to plot the polar curve. The illuminance at any point on the plane is found from eqn [2.9]. The results can be set out as in Table 2.5

Table 2.5

$\theta(°)$	0	20	40	60	80	90
I_θ (cd)	420	450	380	200	70	0
$\cos^3 \theta$	1	0.830	0.450	0.125	0.040	0
E (lx)	67.2	59.8	27.4	4.0	0.5	0
X (m)	0	0.9	2.1	4.3	6.9	—

X is obtained from $X = H \tan \theta$.
E can then be plotted against X.

Isolux diagrams. It can be helpful to display the pattern of illuminance on a specific plane as a series of illuminance contours. Figure 2.13 is an isolux diagram for the data used in the previous example. This type of diagram does provide a method for planning the illumination of a large area such as a car park – its main use being for exterior lighting where there is no reflected light to consider.

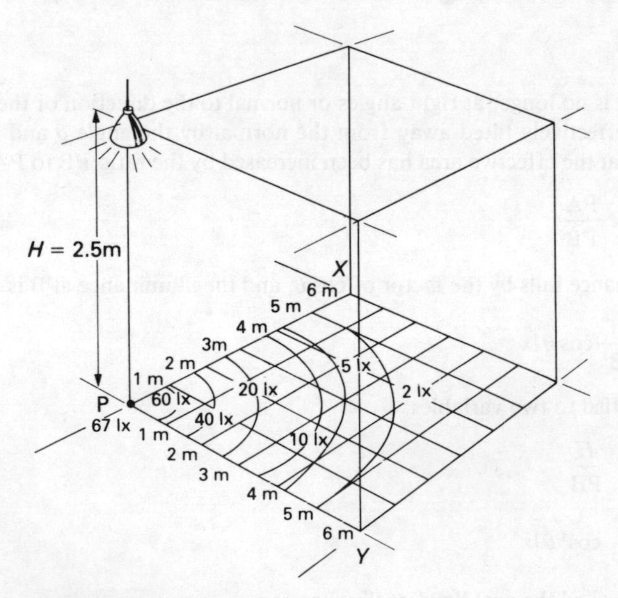

Fig. 2.13 Isolux diagram

Beam or cone illuminance diagrams. A convenient display of direct illuminance at various mounting heights by symetrical luminaires such as recessed spotlights is the beam diagram (also referred to as the cone diagram).

Figure 2.14 illustrates an example for a range of reflector lamps. The vertical scale indicates the height to the plane of measurement and the peak illuminances at the beam centre are given at the side, e.g. 3 m directly below the 100 W lamp the illuminance is 165 lx. Unless otherwise stated, these values refer to the lamp initial lumen output, they are not based on 1000 lm requiring correction.

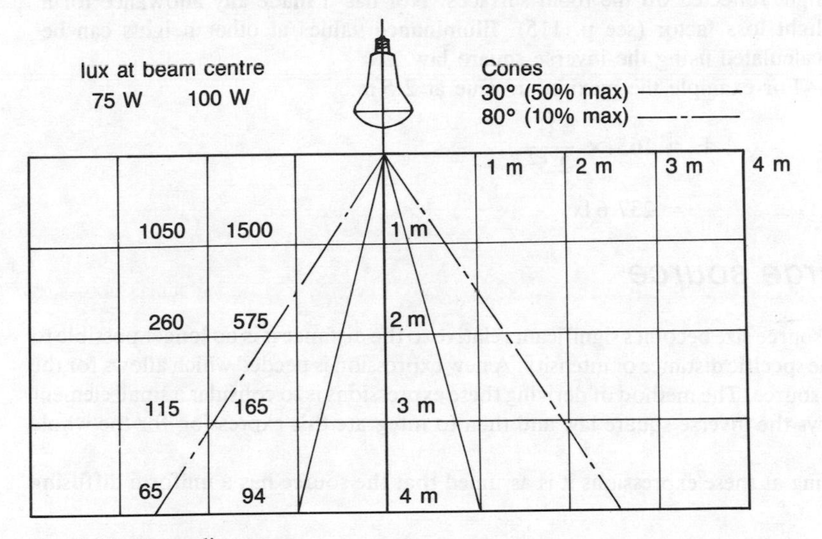

Fig. 2.14 Beam diagram

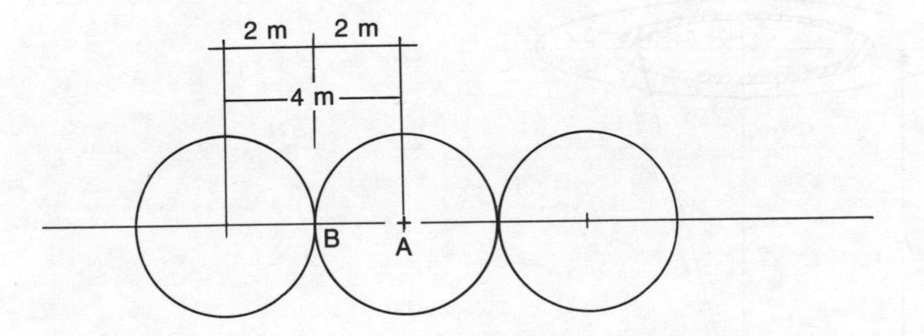

Fig. 2.15 Spacing of 10% beams to produce 5:1 uniformity

There may be two beams shown as in Fig. 2.14. The inner beam is for a maximum : minimum intensity ratio of 2 : 1 and a beam angle of 30°. The larger beam is for a ratio of 10 : 1 and a beam angle of 80°.

To plan an even illuminance the smaller beam is used. However, for display lighting a greater variation can be accepted and the larger beams can be used.

> *Example*
> The 100 W lamp is mounted 3 m above the floor along a corridor. Calculate the spacing and illuminance below a row of lamps to achieve a uniformity ratio of 5 : 1.

Solution
If the 10% beams are drawn to touch each other as in Fig. 2.15 this uniformity will be achieved.

The illuminance at A will be 165 lx
The illuminance at B will be 2 × 16.5 lx, or 33 lx.

This calculation only gives the direct illuminance value and does not include light reflected off the room surfaces. Nor has it made any allowance for a light loss factor (see p. 115). Illuminance values at other heights can be calculated using the inverse square law.

For example the maximum value at 2.5 m:

$$E = 165 \times \frac{3^2}{2.5^2}$$

$$= 237.6\,\text{lx}$$

The large source

When the source size becomes significant relative to the distance it is no longer possible to identify the specific distance or intensity. A new expression is needed which allows for the size of the source. The method of deriving these expressions is to consider a small element which obeys the inverse square law and then to integrate this expression for the whole source.

In arriving at these expressions it is assumed that the source has a uniform diffusing surface.

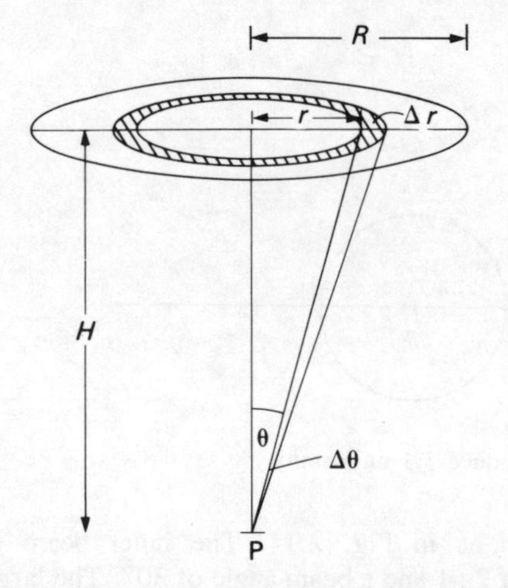

Fig. 2.16 Illuminance from a disc source

The disc source

To find the illuminance on a parallel plane Figure 2.16 shows a uniformly diffusing disc. The illuminance at a point P directly below the centre is:

$$E = \pi L \frac{R^2}{(R^2 + H^2)} \text{lx} \qquad \ldots [2.10]$$

where L is the luminance (cd/m²)
R is the radius
H is the distance.

Derivation. Consider a small ring or radius r and width Δr. This will obey the inverse square law since all parts are equidistant from P. If a diffusing surface has a luminance of L cd/m² then, from eqn [2.3], the intensity in a specified direction is:

$$I = L \times \text{Projected area}$$

e.g. for a flat luminous panel of area S m² and luminance L cd/m² when viewed at angle θ

$$I = L \times S \cos\theta \text{ cd}$$

Considering the elemental ring and the illuminance at P,

$$E = \frac{I}{\text{Distance}^2} \times \cos\theta \text{ lx}$$

To find I:

The area of ring $= \Delta r \cdot 2\pi r$ m²

$\therefore \quad I = \Delta r \cdot 2\pi r \cdot \cos\theta\, L$ cd

Distance $= H \sec\theta$ m

$$\therefore \quad E = \frac{2\pi r\, \Delta r \cdot \cos\theta \cdot \cos\theta}{H^2 \cos^2\theta} \text{lx}$$

This expression contains two variables, θ and r. To integrate put r in terms of θ

$$\tan\theta = \frac{r}{H}$$

$$\therefore \quad \sec^2\theta\, d\theta = \frac{dr}{H}$$

Taking the limiting condition $\Delta r \rightarrow dr$ and substituting r for θ:

$$E = \frac{2\pi L}{H^2} r\, dr \text{ lx}$$

$$= \frac{2\pi L}{H^2} H \tan\theta\, H \sec^2\theta\, d\theta \text{ lx}$$

$$= 2\pi L \sin\theta \cos\theta\, d\theta \text{ lx}$$

$$\therefore \quad \text{Total illuminance} = 2\pi L \int_{\theta=0}^{\theta=\tan^{-1}(R/H)} \sin\theta\cos\theta\, d\theta \text{ lx}$$

$$= 2\pi L \left[\frac{\sin^2\theta}{2}\right]_{\theta=0}^{\theta=\tan^{-1}(R/H)} \text{lx}$$

$$= \pi L \frac{R^2}{R^2 + H^2} \text{lx}$$

Alternatively, this expression could be written

$$E = \pi L \frac{\sin^2 \alpha}{2} \text{ lx}$$

where α is half the angle subtended by the source at P; or

$$E = \pi L \text{ (function } \alpha) \text{ lx}$$

This function of α can be termed the Aspect Factor (AF) as it is a function of the aspect angle. A table can be provided of (function α) against α and this can be used to simplify calculations and to consider surfaces other than uniform diffusing.

Example

If a disc has a luminance of 400 cd/m² and radius of 1.5 m, what is the luminance on a parallel plane 2 m below?

Solution

$$E = \pi \times 400 \times \frac{1.5^2}{1.5^2 + 2^2} \text{ lx}$$

$$= 450 \text{ lx}$$

Approximations. It is possible to use this formula for other shaped area sources. There is a degree of error, but it need not be significant.

Example

Calculate the illuminance 2 m below the centre of a 3 m square luminous ceiling which emits 10 000 lm, given the relationship that

$$\text{Luminance} = \frac{\text{Lumens emitted per m}^2}{\pi}$$

Solution

The area of an equivalent disc is:

$$\pi R^2 = 3 \times 3 \text{ m}^2$$

$$\therefore \quad R^2 = \frac{9}{\pi} \text{ m}^2$$

$$\therefore \quad E = \pi \times \frac{10\,000}{\pi \cdot 9} \cdot \frac{(9/\pi)^2}{(9/\pi)^2 + 2^2}$$

$$= 470 \text{ lx}$$

The linear source

In similar fashion, and using the notation in Fig. 2.17, the illuminance on a parallel plane directly below one end of a uniform diffusing strip source is given by

$$E = \frac{I_0}{2H} (\alpha + \sin \alpha \cos \alpha) \text{ lx} \qquad \ldots [2.11]$$

where I_0 is the downward intensity per metre length of source

α is the aspect angle in radians.

This expression can be derived in a similar way to the disc, but in this case the element is a small transverse strip as shown in Fig. 2.17. It is obtained from the expression

$$E = \frac{I_0}{H} \int \sin^2 \theta \, d\theta \text{ lx}$$

where the limiting value of θ is α, the aspect angle.

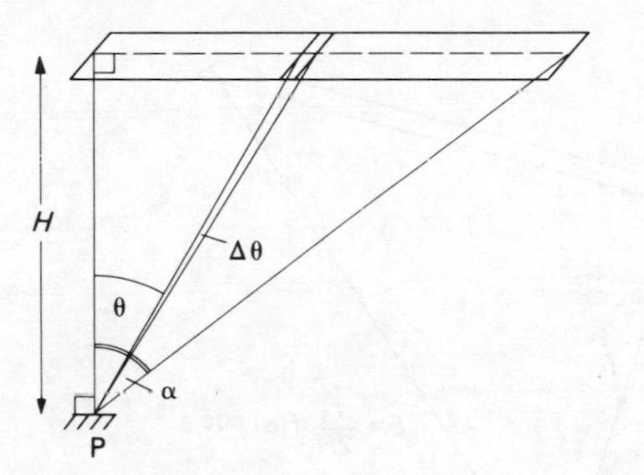

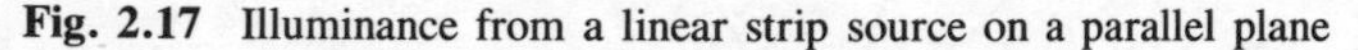

Fig. 2.17 Illuminance from a linear strip source on a parallel plane

The value of E is opposite one end. Where the point lies between the ends, or beyond, the source must be split into separate elements as shown in Fig. 2.18(a), (b).

If the point of measurement is to one side, as in Fig. 2.18(c), then the formula is modified to

$$E = \frac{I_\theta}{2H_1} (\alpha_1 + \sin \alpha_1 \cos \alpha_1) \cos \theta \text{ lx} \quad \ldots [2.12]$$

where α_1 is the new aspect angle
θ is the angular displacement in the transverse plane.

If the source is a flat diffusing strip, then

$$I_\theta = I_0 \cos \theta$$

and the expression would become

$$E = \frac{I_0}{2H} (\alpha_1 + \sin \alpha_1 \cos \alpha_1) \cos^2 \theta \text{ lx} \quad \ldots [2.13]$$

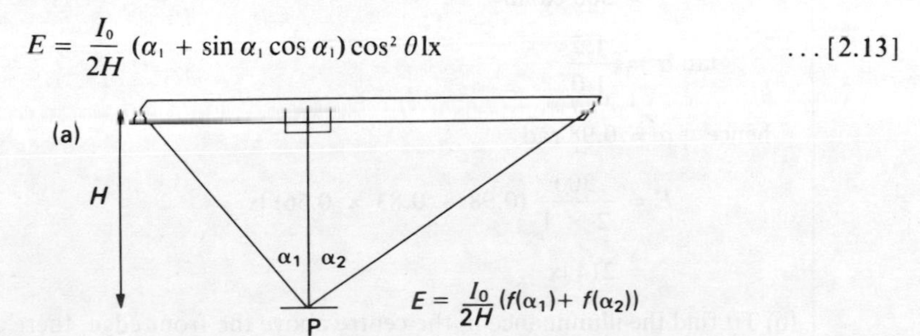

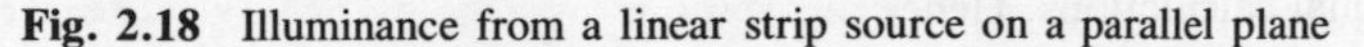

Fig. 2.18 Illuminance from a linear strip source on a parallel plane

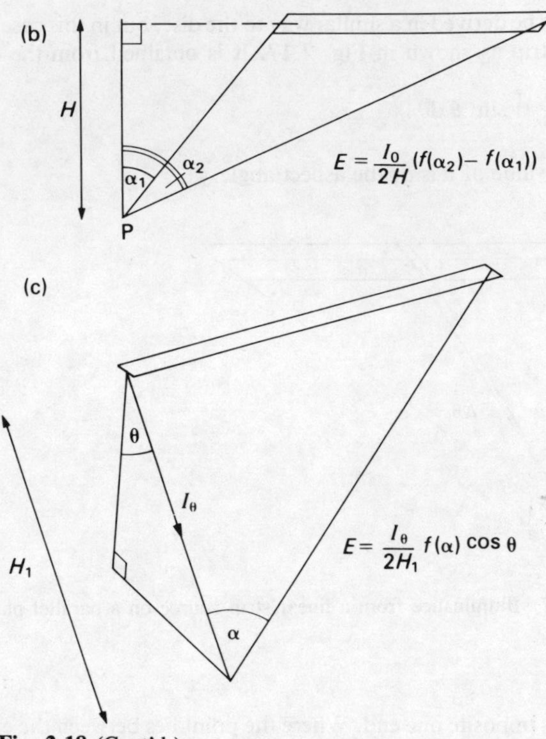

Fig. 2.18 (Cont'd.)

Example

A fluorescent lamp is mounted direct above the front edge of a workbench. Both are 1.5 m long and the mounting height is 1 m. The transverse intensity is 450 cd at all angles.

Solution

(a) The illuminance below one end can be found as follows:

$$I_0 = \frac{450}{1.5} \text{ cd/m}$$

$$= 300 \text{ cd/m}$$

$$\tan \alpha = \frac{1.5}{1.0}$$

$$\text{hence} \quad \alpha = 0.98 \text{ rad}$$

$$E = \frac{300}{2 \times 1} (0.98 + 0.83 \times 0.56) \text{ lx}$$

$$= 214 \text{ lx}$$

(b) To find the illuminance in the centre above the front edge, there are two equal calculations. Hence

$$E = 2 \times \frac{300}{2 \times 1} (0.64 + 0.60 \times 0.80) \text{ lx}$$

$$= 337 \text{ lx}$$

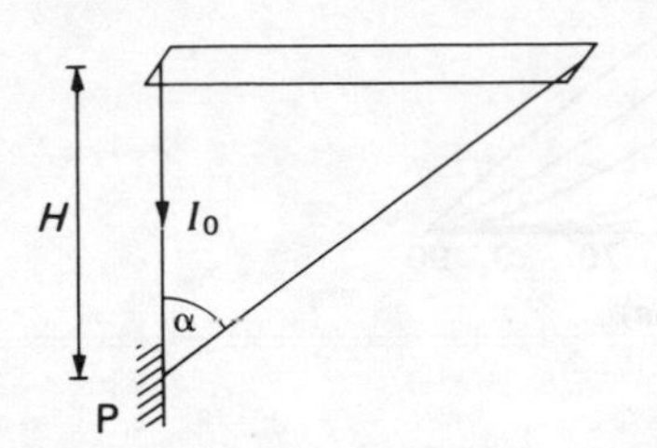

Fig. 2.19 Illuminance from a linear strip source on a perpendicular plane

The above calculations refer to a parallel plane.
If the plane is at right angles, as in Fig. 2.19, then

$$E = \frac{I_0}{2H} \sin^2 \alpha \text{ lx} \qquad \ldots [2.14]$$

Again the calculation refers to a point opposite one end, but this is a far less common calculation.

Use of aspect factor with linear sources. Equation [2.11] can be modified to:

$$E = \frac{I}{2wH} (\text{function } \alpha)$$

where w is the width of the source.

Since $I_0 = L \times \text{Area}$,

$I_0/\text{m} = L \times \text{Width} \times 1$

$$\therefore \quad \frac{E}{L} = K(\text{function } \alpha) \qquad \ldots [2.15]$$

This can be taken a stage further, since although the formula was for a uniformly diffusing strip where $I_\theta = I_0 \cos \theta$, it is extended to include other axial distributions.

Figure 2.20 shows a range of aspect factors for different axial intensity distributions.

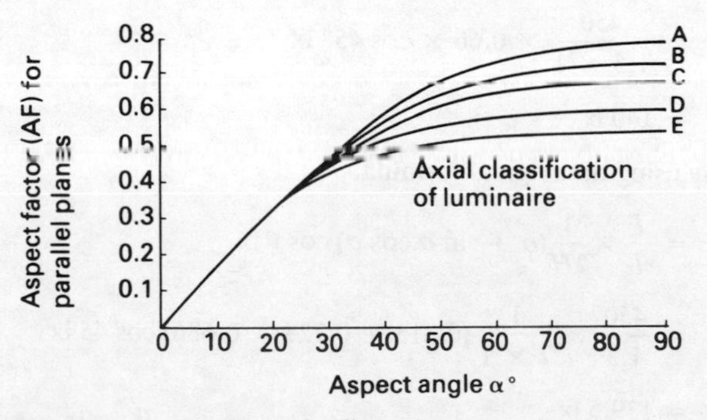

Fig. 2.20 Aspect factor for linear source

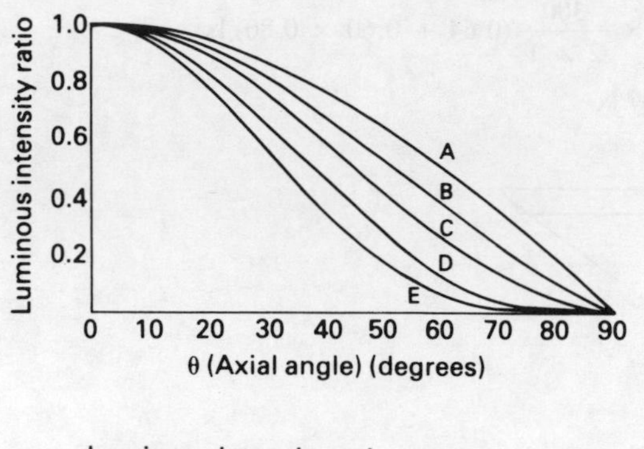

Luminous intensity ratio $= I_\theta / I_o$
where θ is intensity angle in
vertical plane through lamp axis

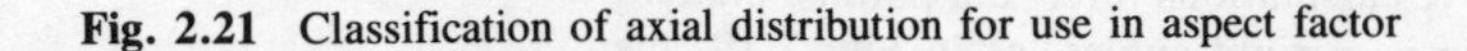

Fig. 2.21 Classification of axial distribution for use in aspect factor

Figure 2.21 shows how a particular axial distribution can be classified. In the case of the examples, the axial distribution is cosine – i.e. that of a uniform diffusing surface – and the classification is A. If there is control of the axial distribution, such as in a prismatic enclosure, the classification will lie between B and E.

Example

Taking the previous example, find the illumination at the far corner of the bench.

Solution

the aspect angle α is given by

$$\alpha = \tan^{-1}\frac{1.5}{2}$$

$$= 46.7^\circ$$

From Fig. 2.18 for classification A, the aspect factor (AF) is 0.66.

$$\therefore \quad E = \frac{I}{lH} \times \text{AF} \times \cos\theta \text{ lx}$$

$$= \frac{450}{1.5 \times 1} \times 0.66 \times \cos 45^\circ \text{ lx}$$

$$= 140 \text{ lx}$$

Checking, by using the original formula,

$$E = \frac{I}{l} \times \frac{1}{2H}(\alpha + \sin\alpha\cos\alpha)\cos\theta \text{ lx}$$

$$= \frac{450}{1.5} \times \frac{1}{2 \times 1}(0.815 + 0.728 \times 0.686)\cos 45 \text{ lx}$$

$$= 139.5 \text{ lx}$$

The slight difference is due to lack of accuracy in finding AF.

The rectangular source

The formulae for rectangular source calculations are not easy to derive and these problems can be solved using tables. The main use of this type of calculation is for windows as these are normally considered to be rectangular diffusing light sources. If the calculation is for side windows, then the plane of illumination is perpendicular to the source; for roof lights the planes will be parallel. Tables for rectangular sources are contained in the Interior Lighting Design Handbook.

Illuminance for non-planar surfaces

The preceding sections have dealt with the calculation of direct illuminance at a point on a flat (or planar) surface. There are a number of situations where this information may be misleading. For example, the illumination of a human face involves vertical and three dimensional surfaces. The horizontal planar value has little relevance.

Two alternatives at present being considered are the spherical surface (scalar illuminance), and the all-round vertical surface (cylindrical illuminance). Methods of calculating and using these values will be found in Chapter 7.

Luminance

This unit has already been partially introduced, it being impossible to discuss without reference to the other units. It expresses the light emitted per unit projected area of light source or reflecting surface. The unit is the candela per square metre (cd/m^2).

There is an alternative non-SI unit, the apostilb (asb). This expresses the light emitted in terms of lumens per unit actual area. It is an easier unit to comprehend since:

$$\text{Luminance (asb)} = \text{Illuminance (lx)} \times \text{Reflectance}$$

There is a simple relationship between the two units

$$\text{Luminance (asb)} = \pi \times \text{Luminance (cd/m}^2) \qquad \ldots [2.16]$$

Concept of perfect diffusing surface

Calculations are usually simpler if reflecting and emitting surfaces can be considered perfectly diffusing. This means the luminance is the same in *all* directions.

Consider a flat diffusing surface of area S m^2:

$$\text{at } \widehat{\theta}\, L = \frac{I_\theta}{\text{Projected area}} \text{ cd/m}^2$$

$$= \frac{I_\theta}{S\cos\theta} \text{ cd/m}^2$$

$$\text{at } \widehat{0}\, L = \frac{I_0}{S} \text{ cd/m}^2$$

$$\therefore \quad \frac{I_\theta}{S\cos\theta} = \frac{I_0}{S} \quad \text{since } L \text{ is constant.}$$

$$\therefore I_\theta = I_0 \cos\theta \ldots$$

This surface can be referred to as a 'cosine diffuser'.

Relationship between lumens and intensity for flat diffuser

In Fig. 2.22 consider a hemispherical shell around a surface S. If S emits ϕ lumens, all these lumens will land on the shell surface.

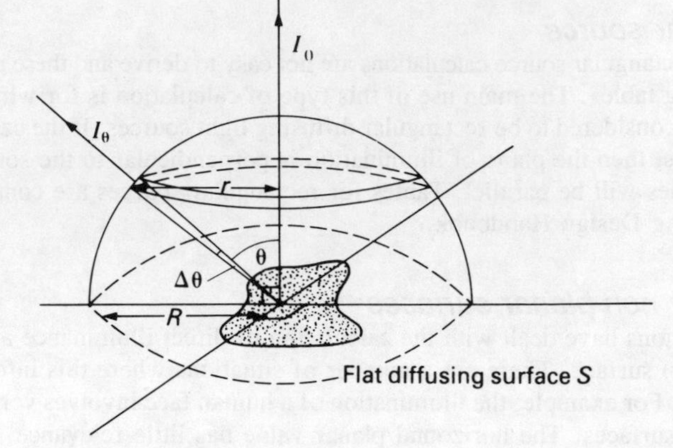

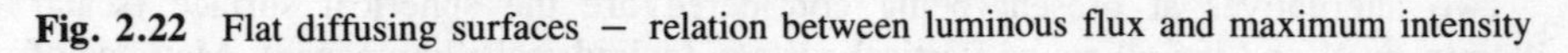
Fig. 2.22 Flat diffusing surfaces – relation between luminous flux and maximum intensity

Consider a small ring of width $R\,\Delta\theta$ and radius r.

$$\text{Area of ring} = \text{Width} \times \text{Circumference}$$
$$= R\Delta\theta\, 2\,\pi\, r$$
$$= R\Delta\theta\, 2\,\pi\, R \sin\theta$$

$\therefore$ Area of ring $= 2\,\pi\,R\sin\theta \times R\,\Delta\theta$

Illuminance on ring $= \dfrac{I_\theta}{R^2}$ lx

$\therefore$ Lumens on ring = Illuminance × Area

$$= \frac{I_\theta}{R^2}\, 2\,\pi\, R^2 \sin\theta\,\Delta\theta \text{ lm}$$

Lumens on hemisphere $= 2\,\pi \int I_\theta \sin\theta\, d\theta$ lm

But, as $I_\theta = I_0 \cos\theta$

Lumens on hemisphere $= 2\,\pi I_0 \int \sin\theta\cos\theta\, d\theta$ lm

$$= 2\,\pi I_0 \left[\frac{\sin^2\theta}{2}\right]_{\theta=0}^{\theta=\pi/2} \text{lm}$$

Applying the limits:

$$\phi = \pi I_0 \text{ lm} \qquad \ldots [2.17]$$

If both sides of [2.17] are divided by the area:

$$\frac{\phi}{S} = \frac{\pi I_0}{S}$$

or

Luminance (asb) = π × Luminance (cd/m^2)

which proves eqn [2.16].

There are three important relationships:

Flat diffuser: Lumens = π × Maximum intensity ... [2.18]

Round diffuser: Lumens = $4\,\pi$ × Average intensity ... [2.19]

Tubular diffuser: Lumens = π^2 × Maximum intensity ... [2.20]

Example
A 1.5 m fluorescent tube has a diameter of 0.038 m and emits 5000 lm. What is its luminance?

Solution
Using eqn [2.20] the maximum intensity is:

$$I = \frac{\text{Lumens}}{\pi^2}\ \text{cd}$$

$$= \frac{5000}{\pi^2}\ \text{cd}$$

$$= 506.7\ \text{cd}$$

The projected area when viewing a tube sideways is that of a rectangle $1.5 \times 0.038\ \text{m}^2$

$$\therefore\ L = \frac{506.7}{1.5 \times 0.038}\ \text{cd/m}^2$$

$$= 8889\ \text{cd/m}^2$$

This chapter has covered the basic calculations of direct luminous flux, and illuminance. Calculations involving the reflected components are considered in Chapter 7.

3 Colour

This chapter deals with a subjective effect. The word 'colour' is meaningless to a blind man and it is difficult to explain in text and monochrome illustration a subject which can only properly be described by personal expression and comparison. The fact that natural adjectives are often used to describe colour illustrates the basic difficulty in attempting to specify colour in technical terms. Even when the technical terms are understood, the description 'blood red' gives an inaccurate but more acceptable impression of a red colour than 'Munsell R 4/14' or CIE 0.6*x*, 0.3*y*.

Coupled with the difficulty in description is the difficulty in expressing the subjective effect. In Chapter 1 the process of colour vision was outlined briefly, and if the energy of different wavelengths produces in the brain the sensation of different colours, then from Table 1.1 it can be deduced that green is a more 'efficient' colour than red in terms of lumens of light. However, red is an arresting colour used to signify danger and green is a 'safe' colour with less visual impact. This contradiction can only be explained by accepting that colours have a psychological as well as a physical effect. It is hard to anticipate the effect colour has on a lighting scheme, and dissatisfaction may arise with a lighting installation, not because there is insufficient light, but because the colour effect of the space or of the room surfaces is disliked.

Mixing of colour

Additive

Isaac Newton first demonstrated and explained the composition of white light, by refracting it through a glass prism into its constituent spectral colours. Colour is an effect of light and if colours are added this implies that different lights are added. The resultant effect on the brain is a new colour, lighter than the originals.

For example, if two monochromatic lights of wavelength 650 nm (red) and 540 nm (green) are mixed and shone onto a white surface, the eye cannot separate the colour of the two sources, but can only recognise the resultant colour mix as yellow. Figure 3.1 illustrates the resultant colour effect of mixing three coloured lights: red, green, and blue.

The red, green, and blue can be called the *primaries* and the resulting yellow, cyan, and magenta the *secondaries*. Any three colours can be used as primary sources provided none of them can be obtained by mixing the other two sources, e.g. no mixtures of green and blue will produce red.

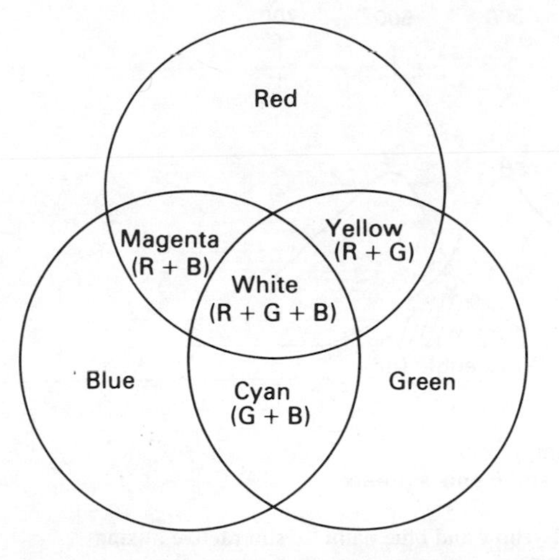

Fig. 3.1 Mixture of three coloured patches of light – additive mixing

Yellow is sometimes referred to as a complementary colour to blue, or as 'minus blue', since the resultant mixture will then contain all three primaries. Similarly, cyan ('minus red') is complementary to red.

The additive mixing of colour is used in many ways; for example:

1. Light sources. The mixing of coloured lights is the basis of stage lighting. Two or three primary colours are used and by suitable dimming control, any desired colour effect can be obtained.
2. Modern lamps. The mixing of colours takes place within the lamp enclosure. For example, the high-pressure mercury lamp in its uncorrected state produces a cold white light, deficient in red. This is partly remedied by introducing a phosphor, yttrium vanadate, which emits red light.
3. Colour television. There are three colour transmitters, each activating a separate pattern of dots on the screen. Enlarged through a magnifier a screen would appear as a mosaic of red, green, and blue dots. At normal size the eye cannot resolve the dots but only their resultant colour mix. A breakdown of the green transmitter would cut out cyan, green, and yellow, leaving a darker picture in shades of blue, magenta, and red.

Subtractive

Most materials which reflect light are selective in what they absorb. A black-body radiator would absorb all incident light and appear black at room temperature. Dark red paint absorbs much of the blue and green, reflecting only red. It has subtracted the blue and green from the light. Similarly, a red filter transmits only the red, subtracting the blue and green. Adding two paints or filters has a cumulative effect, and this is illustrated in Fig. 3.2. Whenever subtractive mixing occurs the resultant colour is darker than the original.

An appreciation of the subtractive effect is essential when relating the colour performance of lamps to the colour of the lamp shades and room surfaces. A deep red decor will appear orange if the lamps used are deficient in red. This cannot be overcome by using a red tinted shade or diffuser around the lamp as this will only reduce the general level of illumination without improving the colour.

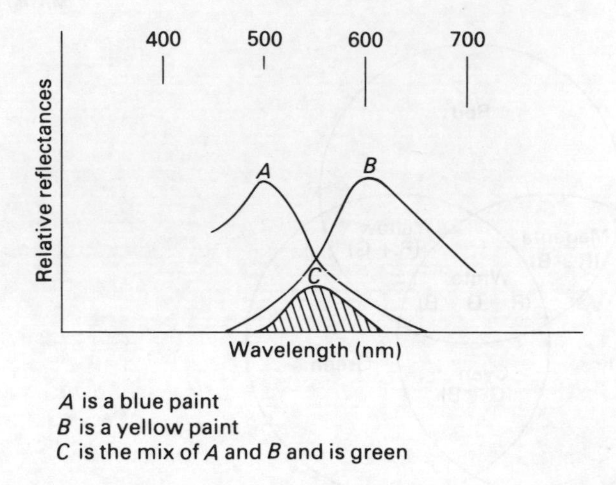

Fig. 3.2 Mixture of yellow and blue paint – subtractive mixing

Specification of colour appearance of surfaces

The colour of surfaces is a combination of the spectral reflecting properties of the surface and the spectral composition of the light source. It may be safest to judge and select colours under filament lamps or daylight but it is quite likely that certain colours will be slightly or even seriously distorted under certain discharge lamp lighting.

The nearest to a universal method of colour specification is the Munsell system.

The Munsell Colour Atlas

Munsell was an American artist and teacher and in 1915 he devised a systematic classification of colour. The Munsell Colour Atlas applies only to coloured surfaces and it is understood and used by many architects and interior designers. Colour is specified in three terms:

1. *Hue* is a description of the actual colour, e.g. red, green.
2. *Value* is a measure of the 'whiteness' of the colour (0 is pure black; 10 is pure white).
3. *Chroma* is a measure of the purity of the colour or 'colourfulness'. Starting with a neutral base, and adding a coloured pigment, the chroma steadily increases until the colour fully saturates the base. Chroma can provide a degree of emphasis, and by choosing colours of different chroma it is possible to vary the emphasis in the pattern of colours. High levels of chroma and value are inevitably associated with yellowish hues, and the lower values and chromas are associated with deep reds and blues. To make full use of the complete range of colours extremes in value and chroma should be avoided other than for specific emphasis.

BS 5252: 1976 Framework for colour co-ordination for building surfaces

There have been a number of BS colour systems, their aim being to try to bring some sanity and unity to the use of colour in buildings. The hope is that the manufacturer can provide a limited consistent range of colour finishes, which the specifier will specify. The benefit to the user is that colours match, despite materials differing, and that his green acrylic bath bears some relation to his green china wash basin, provided both colours are to the same BS number.

BS 5252 is developed from Munsell and colours have three dimensions:

(1) hue – designated by an even number;
(2) greyness – designated by a letter;
(3) weight – designated by a further number.

There are twelve items which have a similar role to 'hue' in the Munsell system:

00 = neutral
02 = red-purple
04 = red
06 = warm orange
08 = cool orange
10 = yellow
12 = green-yellow
14 = green
16 = blue-green
18 = blue
20 – purple-bluc
22 = violet
24 = purple

The *greyness* represents the clarity

A = grey
B = near grey
C = distinct hue
D = nearly clear
E = clear, vivid colour

The *weight* refers to the lightness and relates to both value and chroma in the Munsell system.

Individual colours are identified by a combination of a hue number, a greyness group letter and a weight number, in that order. For example, 12 B 29 means that the hue number is 12, a greenish yellow. The greyness group is B – close to grey, but with a slight hue. The weight number is 29, indicating a darkish tone. The colour is a dark yellow green. The Munsell equivalent is: 2.5GY 2/2

Specification of colour appearance of light sources

Correlated colour temperature

The colour appearance of a 'near white' source can be indicated by its match with a black-body radiator and quoted by the absolute temperature of that radiator. The range is shown in Table 3.1. Examples of lamps and their CCT is given in Table 3. 2.

The most appropriate colour appearance is a matter of personal choice. The only accepted guidelines are that at low lighting levels cool colours tend to give a gloomy appearance and lamps of different colour appearance should not be mixed haphazardly. They can be mixed intentionally as part of the design, e.g. tungsten spotlights can be used for highlights in a space otherwise lit with intermediate colour fluorescent tubes.

Table 3.1 Correlated colour temperature classes and colour rendering groups

Correlated colour temperature (CCT) (K)	*CCT class*
CCT $\leqslant$ 3300	Warm
3300 < CCT $\leqslant$ 5300	Intermediate
5300 < CCT	Cold

Table 3.2 Typical lamp colour properties

Type of lamp	*CCT*	*Class*	*CR group*
Filament GLS	2800	Warm	1A
Fluorescent tube			
Warm white	2800	Warm	3
White	3500	Intermediate	3
Natural	4000	Intermediate	2
Triphosphor	3000	Warm	1B
High-pressure lamps			
MBF		Intermediate	3
HPS		Warm	4
HPS deluxe		Warm	2

The CIE chromaticity diagram (1931)

This is a very precise method of specifying surface colours and light source colours. Its principal use in lighting is for the British Standard specification of fluorescent lamp colours (BS 1853: 1967); it is also used in the pigment and dye industry, and to specify colours of filters.

To understand the system it is necessary first to consider the 'three primaries' method of colorimetry (the measurement of colour). Using three suitable primaries, any other colour effect can be obtained when they are mixed. J. Guild used three monochromatic primaries of blue (435.8 nm), green (546.1 nm), and red (700.0 nm). Figure 3.3 shows how, by mixing these primaries, he could match all the spectral colours. For example, a blue-green of 500 nm wavelength would be matched by 0.54 units of blue and 0.97 units of green minus 0.52 units of red, or

$$C_{500\,\text{nm}} = 0.54\text{B} + 0.97\text{G} - 0.52\text{R}$$

This introduces the concept of negative colours, because using real primaries it is impossible to match a monochromatic colour by only adding other colours since the result will be a lighter colour. Using Munsell terms, it will match in hue but not in chroma.

This method of mixing can be represented graphically using a colour triangle, as shown in Fig. 3.4. RBG is an equilateral triangle whose corners represent the three primaries. The centre of the triangle, W, represents equi-energy white, since it is a colour made up of equal amounts of each primary.

This could be written

$$\text{W} = \tfrac{1}{3}\text{B} + \tfrac{1}{3}\text{G} + \tfrac{1}{3}\text{R} \qquad \ldots [3.1]$$

Numerically $\frac{1}{3}$ = PW = QW = RW and the proportion of a primary is represented by the perpendicular distance of the point from the opposite side of the triangle.

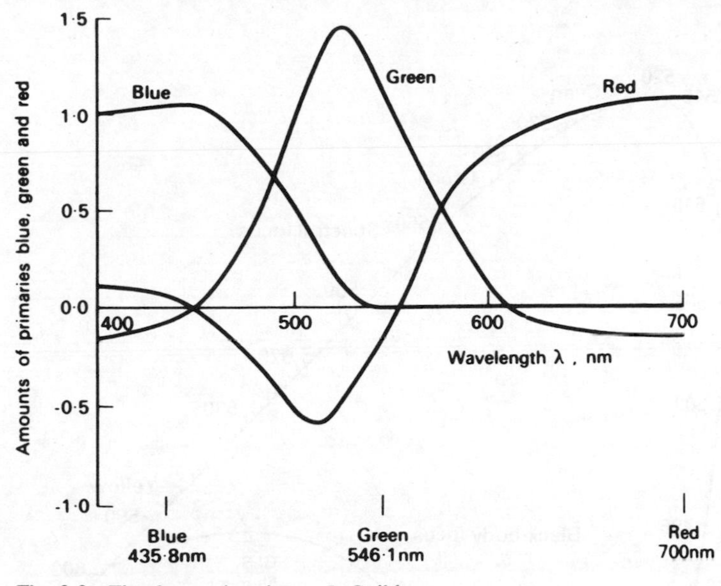

Fig. 3.3 The three primaries – J. Guild

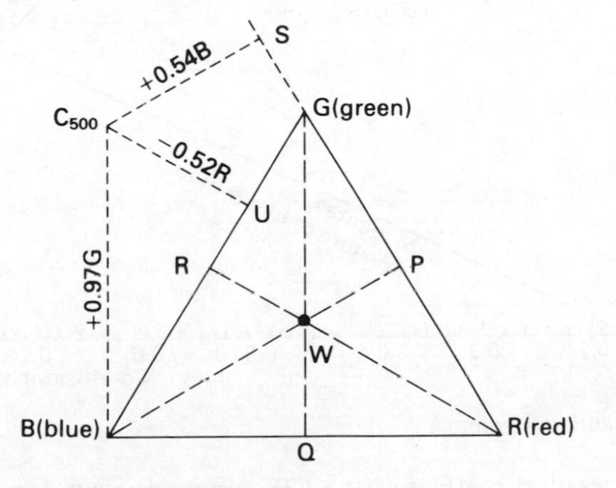

Fig. 3.4 The colour triangle

A colour at point P would be written as:

$$\text{Colour P} = \text{zero B} + \tfrac{1}{2}\text{G} + \tfrac{1}{2}\text{R}$$

the co-ordinates are $(0, \frac{1}{2}, \frac{1}{2})$.

A negative quantity will lie outside the triangle and C_{500} will be shown as in Fig. 3.4. The amount of B is represented by CS, G by CB, and R by CU, where

$$\text{CS: CB: } - \text{ CU} = 0.54{:}0.97{:} - 0.52$$

All the spectral colours (i.e. colours of a single wavelength) – with the exception of 435.8 nm (B), 546.1 nm (G), and 700.0 nm (R) – will lie outside the triangle. These, when plotted and joined, form the spectral curve or locus.

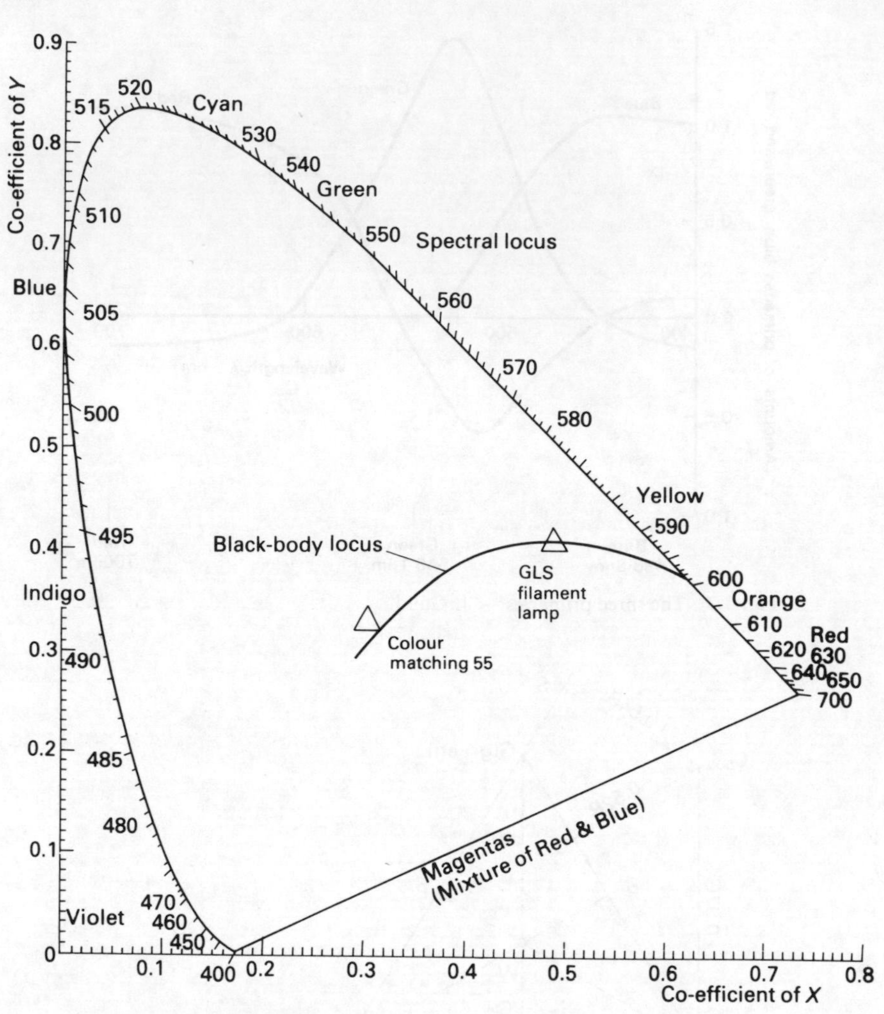

Fig. 3.5 The CIE chromaticity diagram

To overcome this 'negative' coefficient, the CIE system employs another larger triangle which encloses the spectral curve produced by the first triangle using real primaries. This is a mathematical development involving the use of 'unreal' primaries *X*, *Y*, and *Z*. Figure 3.5 shows the CIE triangle.

This triangle is used in similar fashion. The *X* and *Y* co-ordinates *x, y*, are measured off the axes and are called the chromaticity co-ordinates, where $x + y + z = 1$. It is normal to quote only the *x* and *y* values, since $z = 1 - (x + y)$.

All the spectral colours lie on the spectral locus, and all colour mixes lie within the locus.

The black-body locus is a curve representing the colour appearance at various colour temperatures of a full radiator (black body).

There must normally be some tolerance in colour specification and BS 1853: 1960 (amended 1962) quotes this in terms of MacAdam ellipses, which represent a maximum permissible area of deviation in the colour. These are illustrated in Fig. 3.6. This is of particular significance in the measurement of light sources.

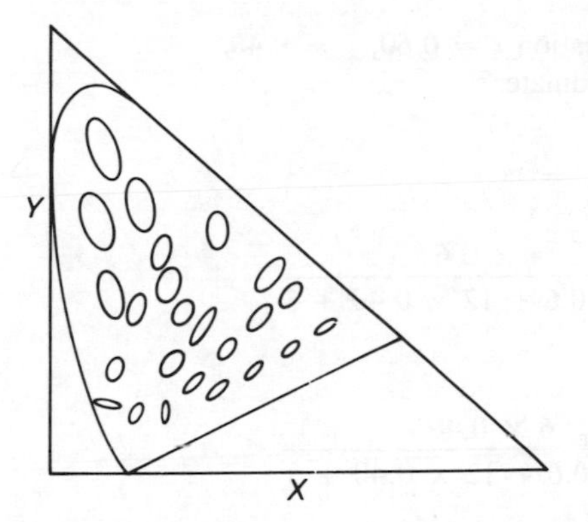

Fig. 3.6 The MacAdam ellipses on the CIE diagram

The Y stimulus is directly proportional to the luminance of the sample colour. This means that if the y co-ordinate for all the spectral colours was plotted, the curve would have the same shape as the eye luminous efficiency curve. The only difference is that at 555.5 nm, $y = \frac{1}{3}$, whereas $V_\lambda = 1$.

The uniformity chromaticity scale (UCS) 1960

The permissible tolerances in the specification of a colour can be indicated by an ellipse (see previous paragraph).

This has not been entirely satisfactory as an ellipse indicates an area of acceptable variation from a specified colour. Figure 3.6 shows that the ellipse varies in size depending on the part of the diagram and colour.

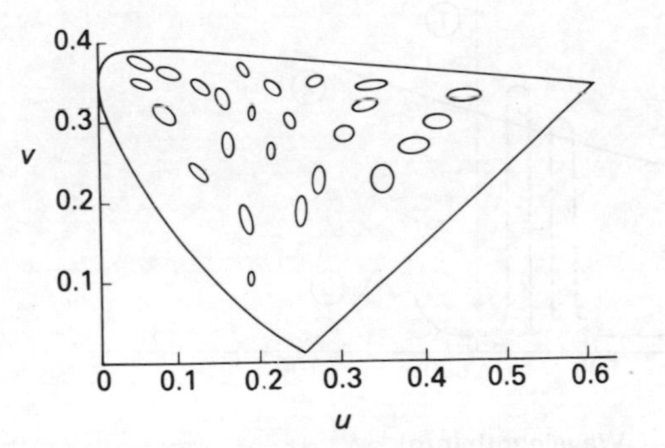

Fig. 3.7 The MacAdam ellipses on the UCS diagram

There have been many attempts to modify the diagram to achieve more even ellipses, and Fig. 3.7 shows the diagram replotted on a $u-v$ axis, where:

$$u = \frac{4x}{-2x + 12y + 3}; \qquad v = \frac{6y}{-2x + 12y + 3} \qquad \ldots [3.2]$$

Example
A colour is red, specification $x = 0.60$, $y = 0.40$.
What are the u, v co-ordinates?

Solution
Using eqn [3.2],

$$u = \frac{4 \times 0.6}{-2 \times 0.6 + 12 \times 0.40 + 3}$$

$$= 0.36$$

$$v = \frac{6 \times 0.40}{-2 \times 0.6 + 12 \times 0.40 + 3}$$

$$= 0.36$$

Colour rendering of light sources

The ability of a lamp to reveal colours is a basic property of its spectrum. Figure 3.8 shows a typical spectrum from three different types of lamps.

① is a low-pressure sodium lamp used mainly for street lighting. The spectrum is line and monochromatic (single wavelength) and the lamp will only reveal that colour. It has no ability to discriminate colours, which is the basis of colour rendering.

② is a filament lamp spectrum. It is continuous, covering all wavelengths fairly evenly with an emphasis on the red end. The colour rendering will be good.

③ is an MBF mercury discharge lamp. Most discharge lamps lie somewhere between ① and ②, and in this case the spectrum is a mixture of some lines, some continuous background spectrum due to the heat of the lamp, and a band of energy in the red end created by the colour correcting phosphor.

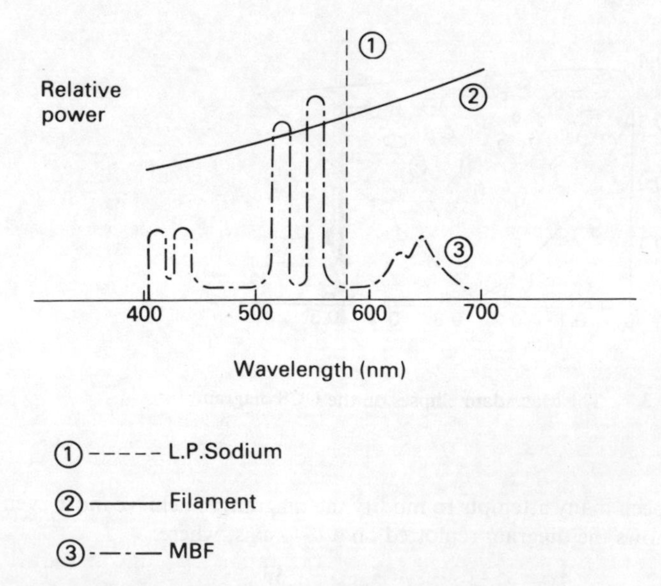

Fig. 3.8 The spectra of various lamps

The CIE colour rendering index

Although there are a number of different colour indices, this is by far the most widely used. The method uses eight Munsell test colours and the general index R_a is based on:

(1) the spectral reflectance of the test colours;
(2) the spectrum of the source under test;
(3) the spectrum of the reference source.

The index has a maximum value of 100. This is for the reference source, which is normally a black-body radiator. The CIBSE 1984 Lighting Code uses a 5-grade scale based on the R_a index.

Table 3.3 CIE colour rendering index

Colour rendering groups	*CIE general colour rendering index (R_a)*	*Typical application*
1A	$R_a \geqslant 90$	Wherever accurate colour matching is required, e.g. colour printing inspection.
1B	$80 \leqslant R_a < 90$	Wherever accurate colour judgements are necessary and/or good colour rendering is required for reasons of appearance, e.g. shops and other commercial premises.
2	$60 \leqslant R_a < 80$	Wherever moderate colour rendering is required.
3	$40 \leqslant R_a < 60$	Wherever colour rendering is of little significance but marked distortion of colour is unacceptable.
4	$20 \leqslant R_a < 40$	Wherever colour rendering is of no importance at all and marked distortion of colour is acceptable.

Although this is the main system used it has its critics and limitations. The main problem is that a single index cannot fully describe the lamp. Two lamps with the same R_a may have widely differing effects on colours and some additional information would be helpful. Sometimes a deliberate distortion of colour occurs, such as with the fluorescent tube 'Deluxe Natural'. It has an enhanced red content which is favoured in the display of cooked and fresh food, but due to the distorted spectrum it only has a relatively low R_a. The CIBSE 1984 Lighting Code includes *both* the colour rendering group and a brief description of the spectrum, as shown in Table 3.4.

Table 3.4

Lamp	*CCT class*	*CR group*	*CR characteristics (visual assessment)*	*Typical application*
Fluorescent tube white	Intermediate	3	Emphasises yellow and to lesser extent, green, subdues red and to some extent blue	Factory, office and shop

(Excerpt from CIBSE 1984 Lighting Code).

Colour constancy

Despite all attempts to specify colour and produce standard sources, the human response to colour is variable and not always predictable.

One example is the ability to recognise that surfaces have the same spectral reflectance, even though their colour appearance is different. This is known as 'colour constancy' and an extreme example could be the lighting of a wall by different coloured spotlights. Provided it is realised that multi-coloured spotlamps are used, no assumption is made that the wall decoration is multi-coloured. There are, however, conditions when an observer cannot be sure if colour difference is due to the source or surface.

Colour matching

If colours are to be matched then it is essential that the process is carried out under controlled conditions. Only too frequently articles that appear to match under fluorescent lighting do not when taken into the daylight. This is called **metamerism** and a non-metameric match is one which matches under any lighting condition. This is only likely to happen if using identical pigments and material.

BS 950 specifies requirements for industrial colour matching, including the performance of the lamp, the construction of the viewing booth, and the illuminance (at least 1000 lx).

4 Lamps

The whole of interior lighting design revolves around the lamp, and development and competition within the industry has resulted in an enormous range of lamps now being available to the lighting designers. The proliferation of lamp types is a mixed blessing because lamp development can often make an existing installation obsolete both in appearance and efficiency before its time. However, a surfeit of choice is not to be scorned. It means that the lighting designer must appreciate the range of lamps at his disposal and keep himself well informed on new developments.

Production of light

The source of electromagnetic radiation has been discussed in Chapter 1. The emission of radiant energy only occurs if the electron also receives energy. The two principal methods are by heating (resulting in thermal radiation) and by collision (passing an electric current through a gas or vapour). Coupled to the collision process is fluorescence, where a phosphor is bombarded by photons as opposed to electrons.

Filament lamps

Temperature radiation

In a tungsten filament lamp the passage of electricity through the filament raises the temperature of the molecules within the filament to the point where they begin to give off light, or they **incandesce**.

This can be explained as follows:

In any body, the molecules which make up its construction are (except at −273 °C, absolute zero), in constant motion. However, the physical state of the material (solid) results in these molecules being formed into fixed patterns or structures. Consequently, their only freedom for movement is vibration about a fixed point. In general terms this vibration process is extremely complex, but the net result is that energy is given to each molecule by interaction and collision. With each vibration some of this imparted energy is released by the molecule in the form of radiation. The effect is analogous to **excitation** (see 'Discharge lamps' below) but in molecular form.

The energy that is radiated obeys 'black-body radiation' laws, and the resulting spectral energy distribution is a function of the temperature of the filament, as shown in Fig. 4.1. The laws are based on a physical concept of a 'black body' or full radiator. This is a

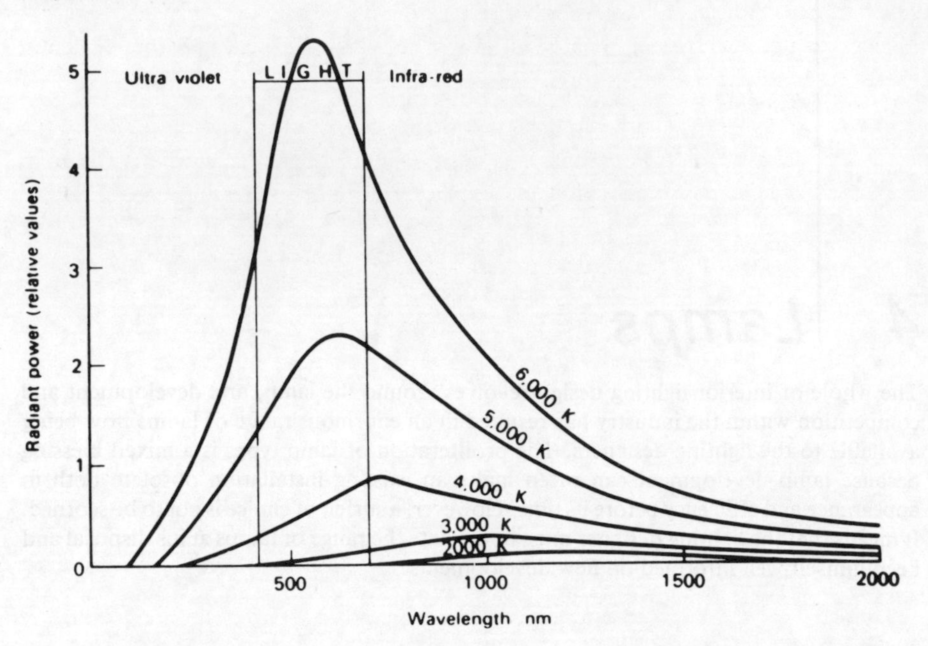

Fig. 4.1 The radiant power from a black-body radiator at different temperatures

body that absorbs all radiation falling upon it. The tungsten filament when heated performs in a similar if not identical pattern to a full radiator.

The share of the spectral energy distribution is given by Planck's equation:

$$P_e = \frac{C_1}{\lambda^5 [\exp (C_2/\lambda T) - 1]} \qquad \ldots [4.1]$$

where P_e is the power radiated (W m^3)
λ is the wavelength (m)
T is the absolute temperature of radiator (K)
$C_1 = 3.7415 \times 10^{-16}$ W m^2
$C_2 = 1.4388 \times 10^{-2}$ m K

In this expression it is seen that the power radiated per unit surface area is dependent only on its temperature.

Two further laws that effect the performance are:

Wien's Radiation Law:

$$P_{e\lambda\max} \propto T^5 \qquad \ldots [4.2]$$

Wien's Displacement Law:

$$\lambda_{P\max} \times T = \text{const} \qquad \ldots [4.3]$$

Summarising: [4.1] gives the spectral distribution at temperature T; [4.2] shows that as T increases, the maximum power increases by the 5th power; and [4.3] shows that as T increases the wavelength of peak radiation decreases. These laws combine to indicate that

(1) the vast majority of the radiant power is in the infra-red;
(2) increasing the temperature not only increases the power output per unit area but also increases the proportion in the visible spectrum.

Filament lamps are efficient radiators of power but very inefficient radiators of light.

Example
If the CCT of the sun is 6000 K, and a filament lamp is 3000 K, assuming the peak radiation from sunlight is at the point where $V_\lambda = 1$, at what wavelength will the peak for the lamp occur?

Solution
V_λ is 1 at 555 nm (see Chapter 1). Therefore, using eqn [4.3],

$$555 \times 6000 = \lambda \times 3000$$
$$= 1110 \text{ nm}$$

The general lighting service lamp (GLS)

A lamp designed for general lighting service (GLS) is illustrated in Fig. 4.2.

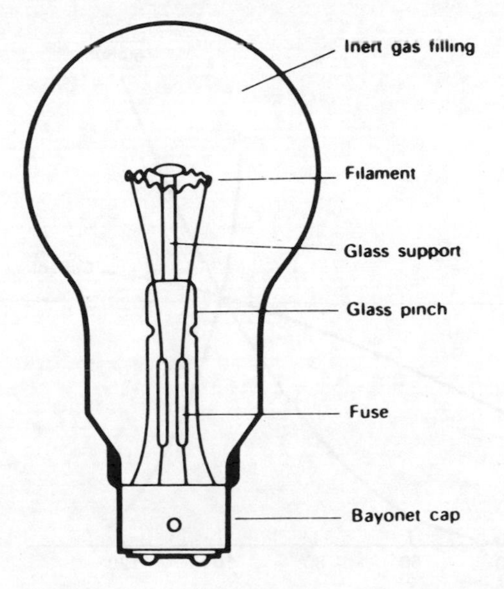

Fig. 4.2 General lighting service – filament lamp (GLS)

Filaments. These are constructed of drawn tungsten wire which is then coiled, and in the lower-wattage gas-filled lamps the coils can be coiled again to reduce convection losses (coiled-coil).

Although the melting point of tungsten is 3600 K, at temperatures above 2800 K the rate of evaporation increases to an extent where the lamp life is drastically reduced. More efficient lamps are available, for example, in projection equipment, but they have much shorter lives.

Gas filling. The rate of filament evaporation can be reduced by raising the vapour pressure in the lamp. This can only be done by introducing gases which are chemically inactive with hot tungsten. An added complication is the loss of heat due to convection and conduction by the filler gas. The most common gas filling is a mixture of argon and nitrogen, and this can only be used with coiled filaments where the heat losses are much less than for a straight-wire filament.

Glass envelopes. These are spherical or 'mushroom' in shape and can be clear, pearl (etched on the inside surface), or 'inside white' (where the inside is coated with silica, titania, or some similar substance). Clear and pearl bulbs have the same efficacy. 'Inside white' bulbs have 4–8 per cent lower efficacy but a much greater degree of diffusion.

Lamp caps. The British GLS lamps normally have Bayonet Caps (BC) up to 150 W. Edison Screw (ES) caps are used for 200 W lamps and Goliath Edison Screw (GES) caps are used for 300–1500 W lamps.

Lamp life. Figure 4.3 shows that small variations in supply voltage cause dramatic changes in life. There are instances where lamps are deliberately over-run to improve their performance. This is true of projector lamps which may only have a life of a few hours in order to achieve much greater temperature and efficacy. Conversely, where lamps are difficult to maintain they may be deliberately under-run to extend their life.

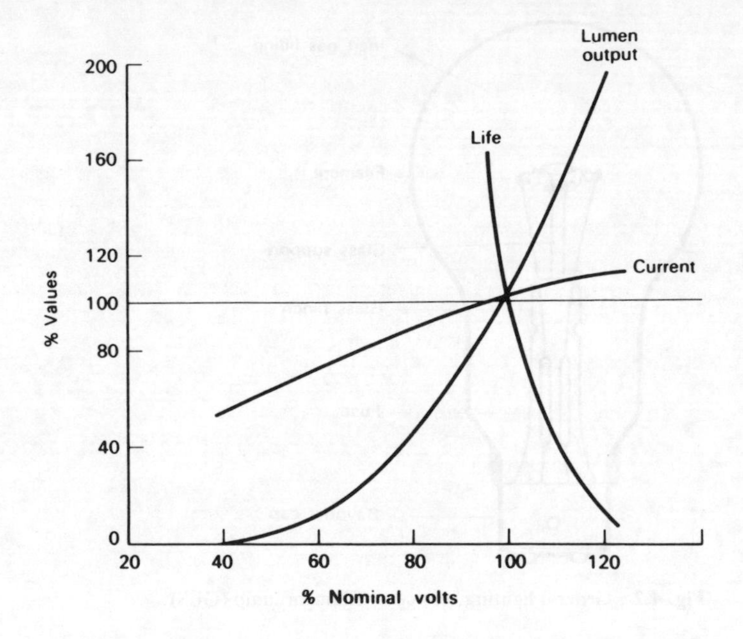

Fig. 4.3 Effect of voltage variation on the performance of GLS lamps

Example
A 100 W GLS lamp operating on 240 V has a light output of 1260 lm and life of 1000 h. What will be its performance on 260 V?

Solution
Referring to Figure 4.3, a 20 V increase represents $\frac{20}{240} \times 100 = 8.3$ per cent. Values can only be obtained approximately from the graph, but 8 per cent will

produce a 40 per cent life and 140 per cent light change. Therefore,

$$\begin{aligned}\text{Life} &= 0.4 \times 1000 \\ &= 400\ \text{h} \\ \text{Light output} &= 1.4 \times 1260 \\ &= 1764\ \text{lm}\end{aligned}$$

Tungsten–halogen lamps

The most important factor deciding the light output of a lamp is the filament temperature, and this determines the life of the lamp.

The addition of a halogen vapour to the gas filling effectively reduces the evaporation rate of the filament. This allows the use of a higher filament temperature and, hence, promotes greater efficacy without the detriment of short lamp life.

The tungsten–halogen regenerative cycle

The halogens are a group of elements comprising fluorine, chlorine, bromine, and iodine which are chemically very reactive with other elements. With the addition of a halogen to the gas filling, a complicated chemical cycle is established whereby the evaporated tungsten particles combine with the halogen to form a re-usable compound called a *halide*.

This compound will condense at temperatures below 250 °C but by constructing the lamp in such a way that the bulb wall temperature is maintained above this value, the tungsten halide is prevented from condensing and will be swept around the inside of the lamp by convection currents. Eventually it will pass near the filament region where the high temperature will cause the compound to dissociate into tungsten, which is deposited back onto the filament, and halogen, which is released to repeat the cycle. Tungsten–halogen lamps therefore do not blacken, and full light output is maintained throughout life.

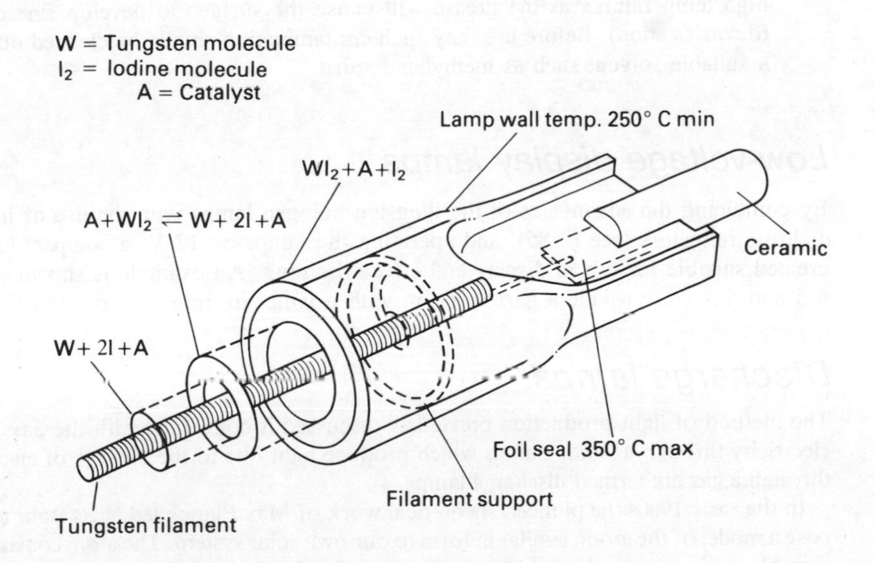

Fig. 4.4 Tungsten – halogen linear lamp

Construction

The bulb must be capable of operating continuously above 250 °C and a form of fused silica called quartz is used. To maintain lamp wall temperatures above 250 °C requires a small bulb size which, in itself, has the added bonus of allowing the use of expensive but efficient Krypton or other rare gases as a filling gas and in allowing the gas-filling pressure to be raised without risk of explosion. Consequently, these lamps operate with higher filament temperatures, resulting in increased efficacy and an increased lamp life, generally 2000 h. A typical linear lamp is illustrated in Figure 4.4.

Operation

There are several operating conditions which must be observed for satisfactory performance.

1. The quartz bulb wall must be maintained above the condensation temperature of the halide (approx. 250 °C).
2. The hermetic seal between the quartz bulb and the molybdenum lead-in wire must be kept below 350 °C. Above this temperature the molybdenum lead-in wire starts to oxidise and a mechanical stress is applied to the constituents of the seal, which breaks.
3. The coolest part of the filament (where it enters the lamp) must be maintained above a critical temperature. Should it not do so, then some corrosion of the filament wire may take place with a reduction of life. For this reason it is not advisable to reduce the voltage to the lamp to less than 90 per cent of its rated value, except where this parameter has been considered by the lamp manufacturer during design.
4. Linear lamps must be operated within 4° of horizontal. Outside this range, the halogen vapour will migrate to the lower end of the lamp, resulting in early lamp failure of the upper part of the filament.
5. Contamination of the outside surface of the lamp wall must be avoided. For example, a deposit of grease by handling will cause the quartz envelope to fail at high temperatures as the grease will cause the surface to develop fine cracks (devitrification). Before use, any such contamination should be cleaned off with a suitable solvent such as methylated spirit.

Low-voltage display lamps

By combining the advantages of the tungsten halogen lamps with the use of integral dichroic reflectors (see p. 85), and operating the lamps on 12 V, a compact lamp is created suitable for use in display and effects lighting. An example is shown in Fig. 4.5 and this will provide a narrow beam with a lamp life in excess of 2000 h.

Discharge lamps

The method of light production previously discussed is concerned with the passage of electricity through a solid. Lamps which produce light due to the passage of electricity through a gas are termed **discharge** lamps.

In the early 1900s the pioneer, theoretical work of Max Planck led Niels Bohr to propose a model of the atom, similar in form to our own solar system. The atom consists of a central, dense, positively charged nucleus surrounded by one or more relatively light, negatively charged particles or electrons. These electrons are attracted towards the

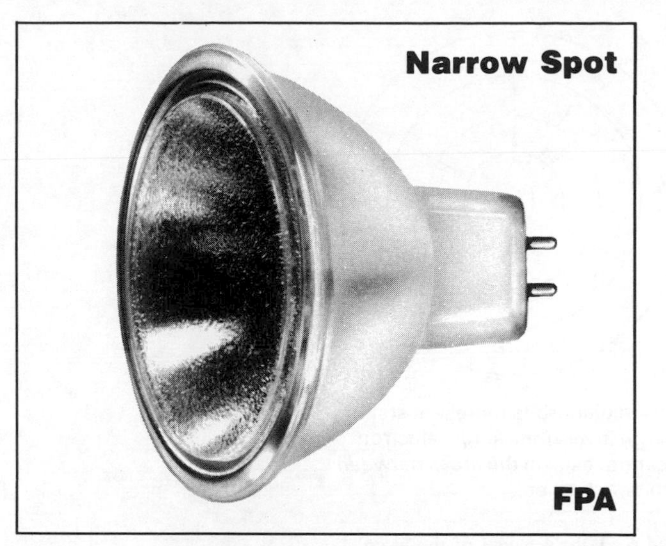

Fig. 4.5 Tungsten halogen lamp with dichroic reflector (courtesy GTE Sylvania)

nucleus but maintain their distance by moving at high speed in fixed orbits around it, in much the same way as our planet orbits the sun.

The electrons moving in the innermost orbits are those most strongly bound to the nucleus; the electrons in outer orbits are more loosely held and are more likely to be involved in the process of collision and subsequent radiation.

Ionisation

Initially when a voltage is applied to a discharge lamp the gas inside the lamp is in an insulating state and will not conduct electricity.

To make the gas conductive a process of ionisation must occur. Normally each gas atom is electrically neutral, i.e. the number of orbiting electrons balances or equals the number of positive charges within the nucleus. By the application of energy to the lamp, outer electrons can be freed so that they become independent negatively charged particles, or **negative ions**. Such a process also leaves the parent atom with a deficiency of one negative charge, so it becomes a **positive ion**. This ionisation process must occur continuously inside the lamp to enable conduction to take place and to replace ions that are lost by recombination.

Ionisation can occur within the lamp in two ways. Initially the application of a high-voltage pulse to the lamp will pull many electrons from their orbits. However, once the lamp is running, charged particles moving along the lamp will collide with gas atoms dislodging electrons in the process.

Excitation and light production

The charged particles move along the lamp because they are attracted towards the electrodes at the end of the lamp.

As they move towards it they accelerate. Many collide with gas atoms before they have gained sufficient energy (speed) to cause ionisation. In these cases the outer electrons are not totally dislodged, but are displaced to a higher orbit.

Bohr showed that displacement can only occur to specific orbit radii and that electrons cannot exist in areas *a*, *b*, *c* or *d* shown in Fig. 4.6.

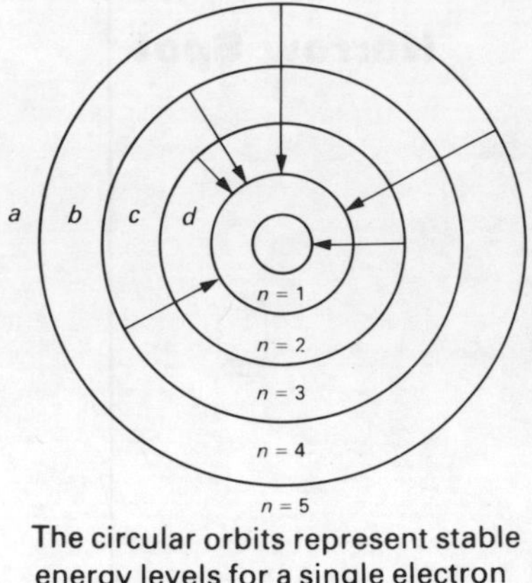

The circular orbits represent stable energy levels for a single electron
It cannot exist in the areas *between* orbits, *a*, *b*, *c*, etc.

Fig. 4.6 Bohr's model of the atom

This displacement is termed **excitation.** An electron may be excited from, say, orbit 2 to orbit 3 or 4 and then return to orbit 2 either directly or from 4 to 3 to 2. At each transition the energy difference between each orbit will be released in the form of photons of radiant power.

High- and low-pressure discharges

In the low-pressure discharge the charged particles move along the lamp a relatively long distance between collisions and will easily build up sufficient speed to cause excitation to the first possible orbit. As the gas pressure is increased the distance between collisions decreases and so the energy gained by the particle during its acceleration time is proportionately lower. However, a study of a practical high-pressure lamp and the power radiated from the discharge shows that the power is due to excitation to higher possible orbits. This may appear to conflict with the basic atomic concept.

The answer is to be found in the collisions which occur before the accelerating particle has had time to build up sufficient speed to cause excitation. In the low-pressure discharge their number is relatively low and has no great effect on the discharge. When the pressure is raised 3000–4000 times, the energy released by each of these collisions produces sufficient heat to raise the temperature of the gas. This high temperature (6000 °C) enables many outer electrons to possess sufficient energy to move to the first possible orbit unaided. Thus the collision processes cause excitation from this point.

Additionally, the increased number of gas atoms present tend to trap any radiation from this first orbit by re-absorbing it and then releasing it many thousands of times as it travels towards the surface of the lamp. The method of producing radiation is basically the same at all pressures, the differences being that at high pressure the atoms combine to form molecules and the energy emissions are at the higher orbit levels.

It is interesting to note that as the first orbit radiations are re-absorbed, the important resonant radiation at low pressure is completely absent at high pressure.

Fluorescence

The low-pressure mercury discharge produces radiant energy within the ultra-violet part of the spectrum. By coating the inside of the arc-tube with a fluorescent coating or phosphor, this UV radiation is converted to visible radiation.

The phosphor must have a strong absorption in the short wave UV band around 253.7 and 185 nm. It must have very low absorption in the visible range and it must have optimum performance at 40–50 °C. Table 4.1 shows a typical range of phosphors and their emission. The colour is also affected by metal activators used with a phosphor.

High-pressure mercury lamps also make use of phosphors, but in this case the excitation energy is at 365 nm and the phosphors are required to produce light only in the red region. Phosphors used include magnesium germanate, and with the improved deluxe lamps, yttrium vanadate.

Table 4.1 Typical fluorescent lamp phosphors

Phosphor	*Colour of light*
Calcium halo-phosphates	White
Magnesium fluoro-germanate	Red
Calcium tungstate	Blue

Gas fillings in discharge lamps

The discharge envelope is filled with a mixture of gases and vapours. The main gas or vapour is the one responsible for the emission of light. This may be in solid state at room temperature, so a further gas is needed to initiate the discharge. When the lamp is running there may be considerable loss of energy due to the current electrons not colliding with the main gas or vapour, and even passing out of the envelope. To increase the probability of collision a buffer gas or vapour is included.

Examples of these three elements are given in Table 4.2.

Table 4.2 Typical gas and vapour fillings

Lamp type	*Main filling*	*Starting*	*Buffer*
Fluorescent tube	Mercury	Argon	Argon
Low-pressure sodium	Sodium	Neon	—
High-pressure sodium	Sodium	Xenon	Mercury
High-pressure mercury	Mercury	Argon and nitrogen	—

Discharge lamp control gear

During starting it is necessary to introduce a higher than normal voltage to the lamp to assist ionisation. However, once the gas has begun to conduct, its resistance will progressively fall (as more and more gas atoms are ionised), until, if unchecked, the mains current will become excessive and the circuit fuse will fail.

To prevent this occurring it is necessary to impose some control on the current that can flow through the circuit. This control equipment or gear could take the form of a large ohmic resistor, but this is wasteful, dissipating considerable amounts of power as heat. A more satisfactory current-limiting device is an inductive resistor, termed a *choke* or *ballast* which is essentially a coil of wire wrapped around a metal core. Current flowing through the wire produces a magnetic field which then hampers the growth of further current thereby 'choking' the current to the desired level.

This type of control only operates on an alternating current supply and in the process creates an overall circuit lagging power factor, where:

$$\text{Circuit watts} = \text{Volts} \times \text{Amps} \times \text{Power factor} \qquad \ldots [4.4]$$

The watts are unaffected, but a low power factor results in an increase in the supply current. In the UK the power factor is normally corrected from 0.5 (uncorrected) to 0.85. It is not normal to correct for lamps below 30 W rating.

All discharge lamps require control gear of some sort from the humblest of neon indicators to the largest stadium floodlight. The size, weight, and cost of such equipment is in proportion to both the lamp wattage and the complexity of lamp technology being used.

Example
If a 65 W fluorescent tube operates on 240 V, 50 Hz supply and the lamp voltage is 100 V, what is the choke inductance necessary to control the lamp circuit?

Solution
The circuit will be as shown in Fig. 4.7. If the lamp power is 65 W and it is assumed to act as a resistor:

$$\text{Power} = V \times A$$
$$65 = 100 \times A$$
$$\therefore \quad A = \frac{65}{100}$$
$$= 0.65 \text{ A}$$

This is the circuit current (I_C).

The choke is a pure inductance and its voltage (V_{Ch}) is 90° out of phase with the lamp voltage (V_L). Therefore,

$$V_{mains} = \sqrt{V_L^2 + V_{Ch}^2}$$
$$240 = \sqrt{100^2 + V_{Ch}^2}$$
$$\therefore \quad V_{Ch} = 218 \text{ V}$$

The impedance of the choke is V_{Ch}/I_C. Therefore,

$$\frac{V_{Ch}}{I_C} = 2\pi \times \text{Frequency} \times L$$

where L is the inductance in henrys.

$$\frac{218}{0.65} = 2\pi \cdot 50 \cdot L$$

hence $L = 1.07$ henrys

Types of discharge lamps and typical control circuits

Tubular fluorescent lamp

This is basically a low-pressure mercury discharge lamp, producing a high proportion of its radiation in the UV region at 253.7 and 185 nm. This is converted to the visible region by the fluorescence of the phosphor coated on the inside of the glass tube. Table 4.3 summarises the energy change for a 'white' fluorescent tube.

Table 4.3 Energy conversion of 65 W fluorescent lamp

Lamp type	*Conducted and convected heat*	*Short wave ultra-violet radiation*	*Visible light*
Plain 65 W lamp, no phosphor	35%	62%	3%
Lamp with phosphor added	35%	16% 30% Infra-red 16%	3%
Total	51%	30%	19%

Lamp construction and performance The low-pressure mercury vapour lamp can have a relatively high light output and efficacy provided the ultra-violet radiation generated in the lamp is converted into light by means of the fluorescent coating. This type of lamp is one of the most common lamps manufactured today. It is used extensively in industrial and commercial premises and is beginning to be used in the home. Its general construction is shown in Fig. 4.7.

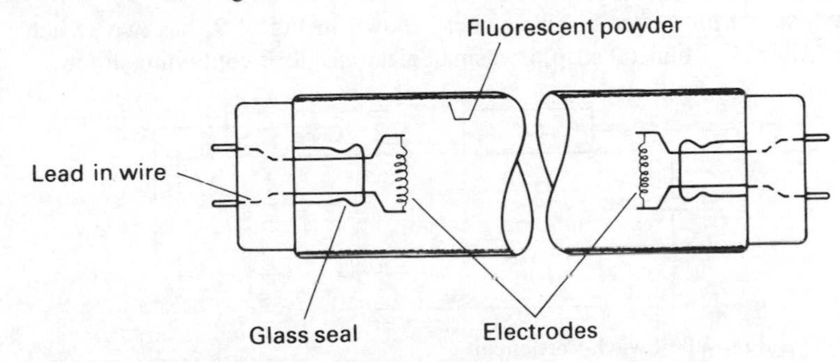

Fig. 4.7 Tubular fluorescent lamp (MCF)

Electrodes The electrodes, which are at the ends of the tube, act as both anode and cathode alternately when connected to alternating current but are normally called cathodes.

Electrons are emitted at the cathode and collected at the anode. For an electron to leave the cathode, it has to be given sufficient energy to overcome the forces holding it to the atoms of the cathode material. This energy is called the 'work function' and different materials have different work functions. The cathode is therefore coated with a thin layer of emissive materials such as the alkaline earth oxides of barium, strontium and calcium, which have a low work function and emit electrons freely on heating.

The main cathode material used is tungsten (because of its low evaporation rates) in the form of a coiled-coil or triple-coil filament or a braided filament, which is coated with the emissive material.

Heating the electrodes by passing a current through them before the arc is struck makes sure that a large number of free electrons exist around the electrodes, making the striking of the lamp easier. During operation the electrodes are kept hot by the passage of the discharge current.

Fluorescent coating (phosphor) The amount of light emitted depends upon the combination of phosphors used. The wavelengths radiated by the phosphor vary with the chemicals used. For example, the phosphors used in present general-purpose fluorescent lamps are **halo-phosphates** which contain calcium, antimony, chlorine, fluorine, and manganese. With no manganese the colour of the lamp is blue and, by adding different amounts of manganese, blue-white and white to yellow-white lamps are obtained. A range of 'white' lamps is produced from 'colour matching' to 'warm white', each having a different colour appearance and colour rendering properties.

One important effect of the choice of phosphors for a given colour appearance is that, in general, the better the colour rendering, the lower the light output. Since the introduction of new generation 'hexagonal aluminate' phosphors it has been possible to provide both high efficiency light output and deluxe colour rendering with (eventually) a variety of colour appearances. This is achieved by using a new group of phosphors having selected high-energy spectral band emissions at three critical frequencies which correspond to high sensitivity eye colour reception mechanisms (having high V_λ values).

Control gear In fluorescent lamp circuits operated on a.c. supplies, the ballast is usually a choke and is connected in series with the lamp, as shown in Fig. 4.8. There are many more complex circuits used to control the lamp current and to assist in striking the lamp, and the term *ballast* is used to cover all forms of control gear.

Switch-start circuits

Glow type starter switch This switch, shown in Fig. 4.9, has two switch contacts, one of which is a bimetal strip, in a small glass envelope containing argon.

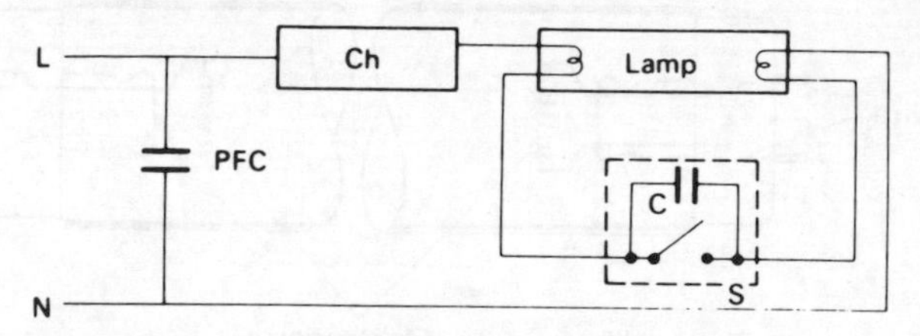

Fig. 4.8 Switch-start circuit

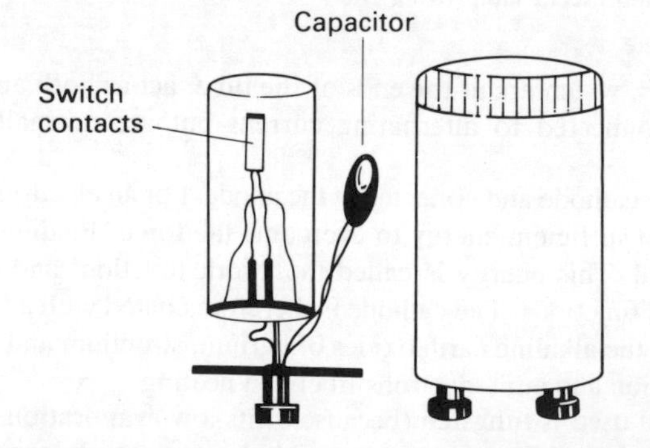

Fig. 4.9 Glow type starter switch

When the supply is switched on the bimetal strip is cool and the contacts are open. Current flows through the choke, through the first electrode, through the argon in the starter switch (in the form of an arc), through the second electrode, and back to the supply. The arc in the argon heats the bimetal strip causing it to bend towards the other contact of the starter until the two contacts close. Current still flows through the two electrodes and the starter switch contacts, and the electrodes are pre-heated, liberating electrons into the arc tube. The argon arc no longer operates and as the source of heat is removed the bimetal strip cools and begins to straighten until the contacts open again.

The opening of the switch causes the lamp arc to be struck because a voltage pulse is induced in the choke and because of the pre-heating of the electrodes. When the arc is struck the voltage across the lamp falls to about its stabilised value.

As the starter switch is connected across the lamp, the voltage across the two contacts is also reduced and is not enough to cause an arc in the starter switch argon gas, and so they remain open until the lamp is required to be lit again. If the lamp does not start the starter operates again until the lamp eventually strikes or the supply is switched off.

Starter capacitor The small capacitor which is connected across the starter terminals (i.e. in parallel with the switch contacts) reduces emission by the lamp at radio frequencies and also improves the waveform of the starting voltage pulse.

Circuit operation on starting Having considered the sequence of events that takes place in the starter switch, this must be related to the lamp and choke during this time. First there is a current flowing through the choke, cathodes, and starter switch. This current is about 1.5 times the running current; it heats the cathodes and causes free electrons to be emitted and ionisation to take place near the electrodes. When the starter switch opens, the choke opposes the change by producing a voltage which is in the same direction as the supply voltage and therefore adds to it. A high-voltage surge across the lamp is enough to cause an arc between the electrodes of the lamp.

Electronic starter switch. A switch is available, at extra cost, which uses a thyristor trigger. This sets up a pre-heat electrode circuit and a high voltage is generated. Starting is near instant and, in the event of lamp failure, the electrode current ceases. This overcomes damage to the starter and ballast.

Starterless circuits

Although the switch-start circuit is reliable and is more suitable for low-temperature applications, the starter switch is the one item in the circuit with moving parts liable to fail. Also, in some applications its operation can be a nuisance due to flicker until the lamp has struck – especially with old lamps.

Starterless circuits operate without a starter switch and so reduce the maintenance required. They have disadvantages, however: they are more complex and expensive; ballast losses are higher than those of the simple choke; and they are more susceptible to difficulties in starting at temperatures below 5 °C. The circuit is shown in Fig. 4.10.

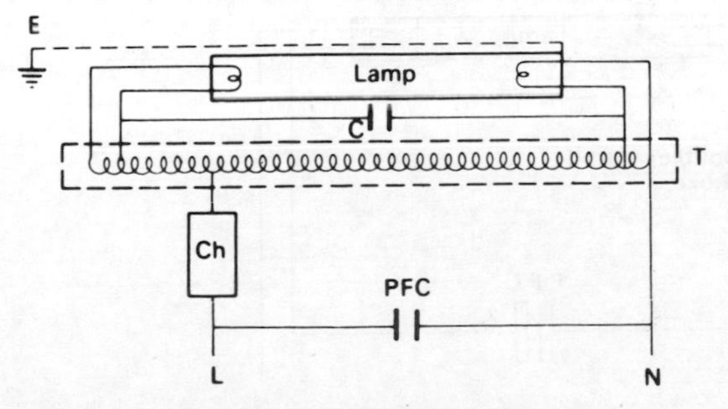

Fig. 4.10 Starterless transformer circuit

The major change from the switch-start circuit is that a transformer is used to provide a low voltage across each electrode, which is contained within the ballast canister together with the choke.

When the circuit is first switched on, nearly all the supply voltage appears across the lamp and the transformer. Current flows through the electrodes from the two transformer windings, one heating each electrode and producing emission of electrons, until there are enough free electrons to enable the lamp to strike with only the supply voltage across the lamp. Once the arc has struck the current through the choke increases and therefore the voltage drop previously across the transformer now appears across the choke, reducing the lamp and transformer voltage so that the lamp current is limited to its correct value. The voltage across the electrodes falls to about half.

This circuit is more complex and expensive than the switch start but is the only circuit suitable for use with dimming control. To dim fluorescent tubes the lamp voltage and filament temperatures must be maintained while the lamp current is reduced. A typical circuit is shown in Fig. 4.11.

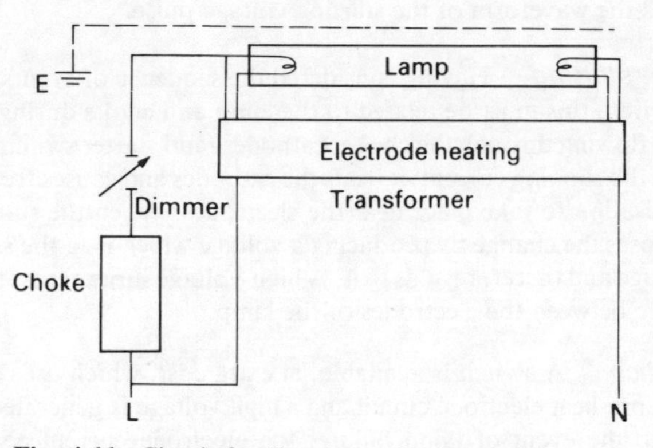

Fig. 4.11 Circuit for dimming a fluorescent tube

Semi-resonant start circuit

This comes within the category of starterless circuits, but is simpler and has lower gear losses than the transformer type. It comprises a double wound choke with two equal coils wound in opposition and a series capacitor. The term *resonant* is used since the subcircuit of the lamp, coil B, and the capacitor (Fig. 4.12) produces a high lamp voltage and filament current for starting (the capacitor and coil combination have minimum impedance). When the lamp is running, the coil B becomes ineffective and the circuit is coil A, lamp, and capacitor. This circuit will operate at low temperatures.

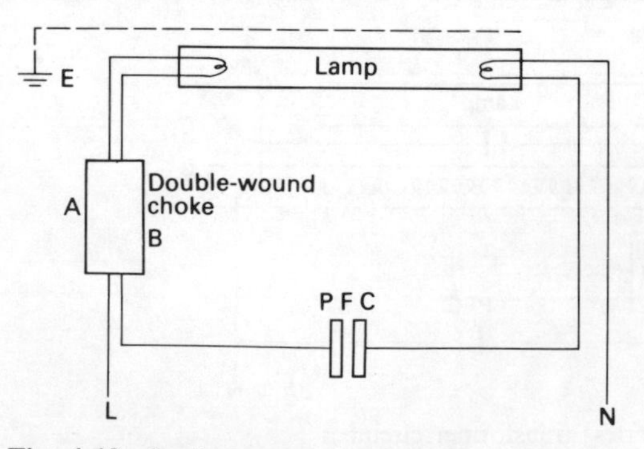

Fig. 4.12 Semi-resonant start circuit

Electronic control circuits

Raising the electrical frequency to the lamp from 50 Hz to a high frequency of around 30 kHz brings a number of significant improvements to performance. These include:

reduction in ballast losses, and weight;
elimination of flicker and noise;
improved lamp efficacy and life;
instant lamp start.

The electronic element of the circuit is the frequency conversion. Figure 4.13 shows an example of this type of circuit.

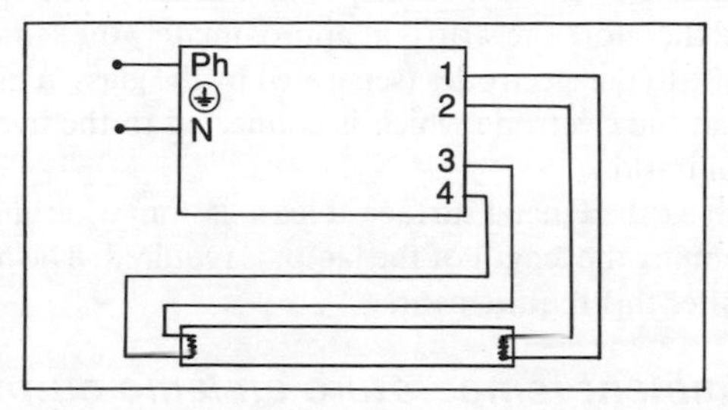

Fig. 4.13 Typical circuit for a high frequency electronic ballast

The total consumption of the HF lamp and ballast is considerably less than that of conventional fluorescent lighting, typically saving 30 per cent of energy cost for equivalent light output. Even more savings are available for air conditioned buildings, where a smaller capacity plant could be installed owing to the reduced heat load from the HF lighting.

The HF system offers particular advantages in twin-lamp luminaires, because a single ballast drives two lamps instead of the two separate ballasts in conventional luminaires.

Table 4.4 compares HF energy effectiveness against conventional lamps in a widely used luminaire type. This highlights the halving of control gear losses, and the big jump in light output versus power consumed by the complete lamp/luminaire system.

Table 4.4 Comparison of high frequency with conventional circuits

	Conventional luminaire with 38mm MCF lamps	*HF lamps and luminaires 26 mm lamps*
Lamp wattage	2 × 65W	2 × 50W
Control gear losses	2 × 14W = 28W	1 × 11W = 11W
Total consumption	158W (taken as 100%)	111W (saving 29%)
Light output	2 × 4900 = 9800 lumens	2 × 5100 = 10.200 lumens
Overall efficacy	62 lm/W	92 lm/W

Lamps for starterless circuits

The starterless circuit does not produce a high-voltage pulse to strike the lamp, and so lamps have been developed which can start, after a short period of pre-heating, on the normal mains voltage. These lamps may also be used on switch-start circuits.

There are two types designed for the starterless circuits. One (MCFA) has a thin metal strip running the length of the tube and connected at each end to the metal lamp caps which must be earthed. The other (MCFE) has a silicone coating over the outside of the lamp. It has been found that the voltage required to start a lamp varies with the lamp surface resistance and that the voltage required is at a minimum if the resistance is *very high* or *very low*. The metal strip provides a very low resistance; the silicone coating a very high resistance. A further advantage of an earthed strip is that usually the neutral of the supply is connected to earth and therefore the strip is at approximately the same potential as the neutral. As the strip is close to the electrodes (separated by the glass) a high voltage now exists between the strip and the electrode which is connected to the live terminal. This high potential assists in ionisation.

To assist the silicone lamp, an earthed metal surface at least 25 mm wide and less than 10 mm away from the lamp, running the length of the lamp, is required. The metal spine of the luminaire normally satisfies this requirement.

Effect of changes in ambient temperature on lamp output and starting

If the air temperature around the lamp is high, the pressure of the mercury vapour inside the lamp rises and the light output of the lamp decreases. If, on the other hand, the air temperature is low, the vapour pressure will not be high enough for sufficient collisions to take place between the free electrons and the mercury atoms, and the light output goes down, as shown in Fig. 4.14.

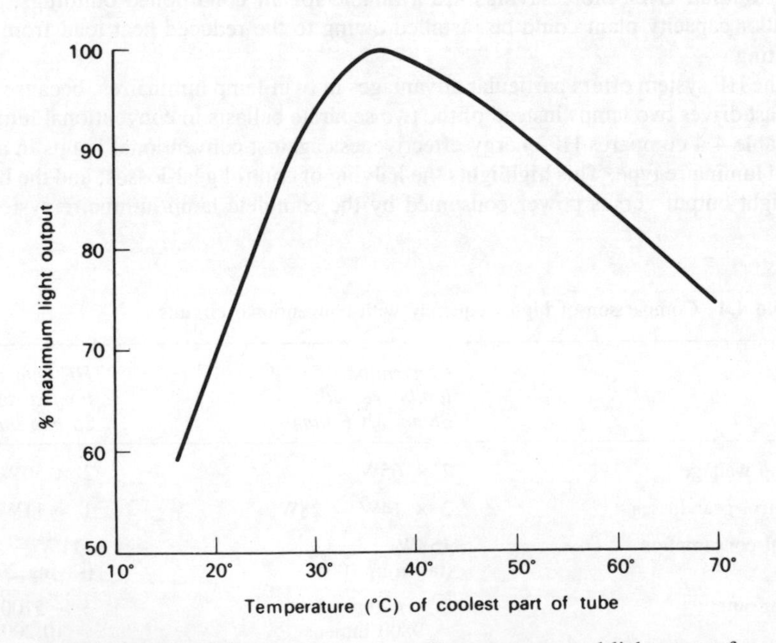

Fig. 4.14 The relation between tube wall temperature and light output for a 65 W MCF lamp

There is a narrow band of temperature at which the light output is a maximum. Normal fluorescent lamps are designed to run in an air temperature of about 25 °C, with a lamp wall temperature of 40 °C which produces a vapour pressure of about 800 Pa. When lamps are operated above or below this temperature the light output will be less than the declared value.

The vapour pressure inside the lamp can be reduced by simply cooling a single area of the lamp, because the excess mercury will condense at the cool spot. For example, heat can be conducted away from the lamp by placing a metal shoe in contact with the outer surface of the lamp. The shoe has to be held in position by a device holding it to the luminaire body (through a hole) and it must be spring loaded to maintain adequate contact. Various other methods have been used, such as setting the electrodes well into the lamp or fixing an insulating shield between the electrodes and the lamp cap to keep the ends of the lamp cool, but all these methods have been superseded by the *amalgam* (TLH) lamp.

In the amalgam lamp, cadmium or indium is inserted in the lamp either on the electrode mount or in the form of a separate ring on the inside of the glass envelope. The indium combines with the mercury in the lamp, forming an amalgam of mercury and indium. The amalgam stabilises the mercury vapour pressure at high temperatures and maintains a high light output.

Effect of changes in supply voltage

The changes in lamp volts, light output, and lamp current that occur with changes in supply voltage are shown (for the switch-start circuit) in Fig. 4.15.

The changes are not as drastic as those produced by variation of the voltage of a tungsten lamp, e.g. a 1 per cent change in voltage produces only a 1 per cent change in light output.

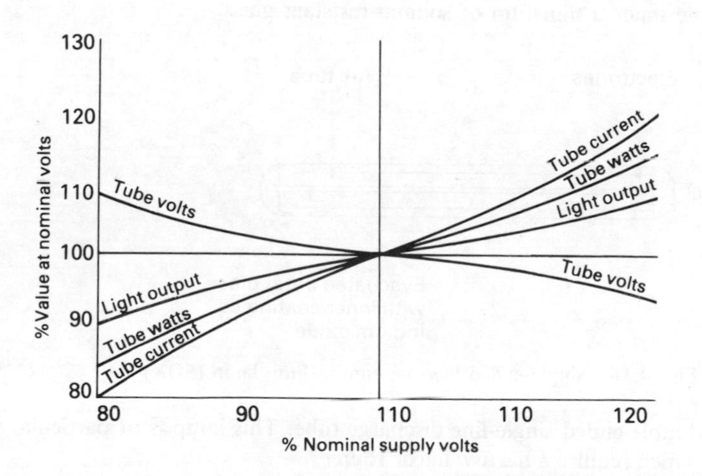

Fig. 4.15 Effect of voltage variation on the performance of a 40 W MCF lamp

Effect of change in supply frequency

The normal changes of ± 1 per cent in supply frequency have about the same effect on inductive circuits as a 1 per cent voltage change. Operating the lamp on a normal supply other than 50 Hz, say 60 Hz, has no effect on the lamp characteristics provided the control gear is designed for the frequency used. Operating the lamp at frequencies above

400 Hz improves the light output, and reduces the fluctuations in light output, the size of the control gear, and the losses. These improvements are quite significant, and designing control gear to operate at 400 Hz and above is worthwhile when such a supply is available – e.g. in aircraft, or for operating from low-voltage supplies such as in caravans and boats. When connecting to normal 50 Hz system the cost of changing the frequency outweighs the advantages gained.

Low-pressure sodium vapour lamp

Light is produced by the action of the excited electrons returning to lower energy levels. The sodium vapour arc produces a number of spectral lines but most of the visible radiation (95 per cent) is concentrated in a very narrow region of the spectrum at 589.0 and 589.6 nm (known as the sodium 'D' lines). The light output is therefore *monochromatic* (single coloured) and the lamp is easily recognised by its yellow appearance.

Although this yellow light has the disadvantage that colours are greatly distorted, it has the advantage that, as the relative response of the eye (V_λ) is high at 589 nm (approximately 0.75), it has a potential for a high lumen output, depending upon how much energy is actually put into the lamp.

It is now possible to manufacture low-pressure sodium lamps with efficacies of about 200 lm/W, but not all of the lamps produced are this efficient and the efficacy varies with the type of lamp and lamp wattage.

Lamp construction and performance

There are a number of different types of lamp. The two main types are SOX and SL1.

SOX is a single-ended lamp with an integral outer jacket coated with a heated reflector (see Fig. 4.16). The inner discharge tube is made of 2-layer glass, the outer being soft glass and the inner a thin film of sodium-resistant glass.

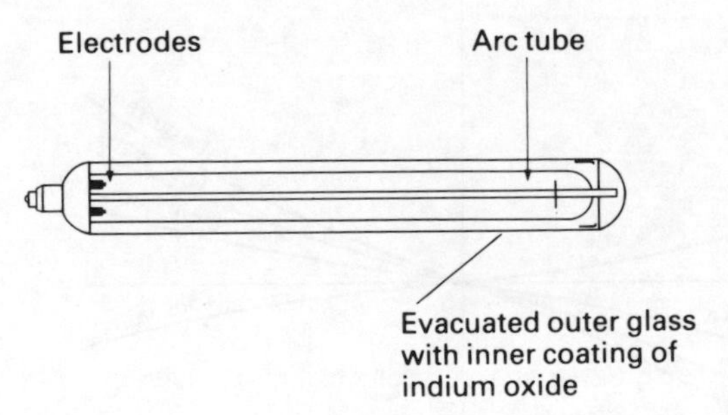

Fig. 4.16 Single-ended low-pressure sodium lamp (SOX)

SLI is a double-ended single-line discharge tube. This lamp is of particular value in floodlights which require a narrow linear source.

The electrical characteristics are similar to those of the fluorescent tube. The lamps are not temperature sensitive as they operate within a vacuum jacket.

Lamp circuits

At room temperature the metallic sodium is solid, and in order to vaporise the sodium, neon and argon are added. Initially a neon–argon arc is formed and the heat produced gradually vaporises the sodium. The lamp has a run-up period of about 10 min before maximum light output is reached.

A transformer is used in the sodium lamp circuit because of the 'higher-than-mains' ignition voltage required, i.e. approximately 470 to 490 V. If a normal transformer is used, a choke would also be needed to limit the current after the discharge has struck because of the negative resistance characteristic of the arc. Such a circuit is, however, expensive in material usage, when one considers that the transformer output voltage is about 500 V while the normal operating lamp voltage required is 100 V.

A lot of material can be saved by, first of all, using an **auto**-transformer and, secondly, by making it an **auto-leak**-transformer which removes the need for a separate choke (see Fig. 4.17).

This leak transformer type of circuit has a low power factor (0.3) and a capacitor is connected across the supply to increase the power factor to an acceptable value.

The **linear** lamp can be operated on circuits similar to fluorescent lamps.

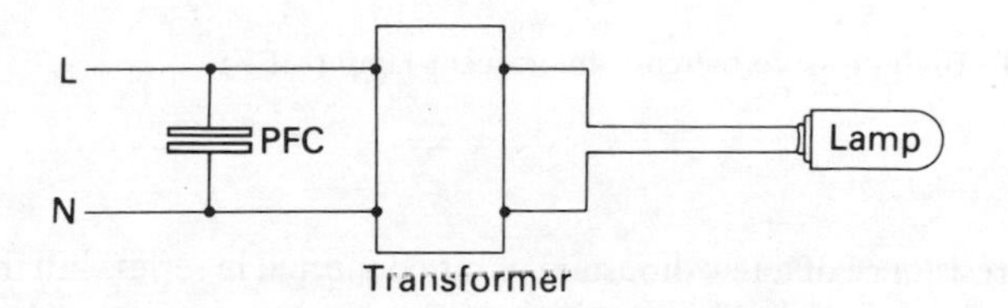

Fig. 4.17 'Leakage-flux' transformer circuit

High-pressure mercury lamp (MBF)

Increasing the vapour pressure of a discharge has the effect of changing the spectral power distribution by increasing the width of the spectral lines (bands) produced, and by changing the relative intensities. Absorption of the resonance radiation also takes place.

Lamp construction and performance

The basic high-pressure mercury vapour lamp shown in Fig. 4.18 consists of a glass outer envelope containing the quartz arc tube, its supports, and the lead-in wires. This outer envelope contains nitrogen to dissipate the heat produced and maintain the arc tube at the correct temperature.

Inside the arc tube are a small quantity of mercury, argon (which acts as a starting gas), and the tungsten electrodes. There are usually three or four electrodes - two of which are main electrodes and a smaller auxiliary electrode (or electrodes) used to initiate the arc. The auxiliary electrode is very close to a main electrode and is connected through a resistor to the electrode at the other end of the lamp. At switch-on the full mains voltage exists between the auxiliary electrode and its adjacent main electrode. To assist in electron emission, the main electrodes are impregnated with a low work function emissive material.

Lamp operation

At room temperature the mercury in the lamp is liquid and the argon is included in the arc tube to start the arc and to evaporate the mercury. When the lamp is connected to the supply, there will initially be no current flow and mains voltage will exist between the two main electrodes and between a main electrode and its auxiliary electrode via the internal resistor. This voltage will produce a local arc in the argon gas with the current limited to a

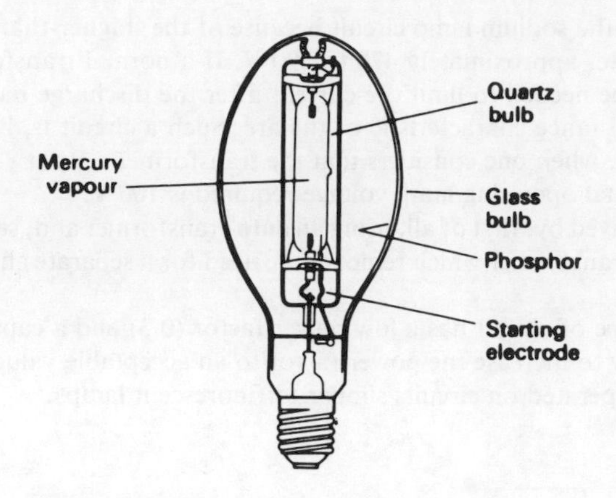

Fig. 4.18 High-pressure mercury fluorescent lamp (MBF)

safe value by a high resistance of a few thousand ohms connected in series with the auxiliary electrode as shown on Fig. 4.18. The initial arc produces ionisation, and an argon arc is soon struck between the two main electrodes. This produces further ionisation until sufficient mercury ions exist for the mercury arc to be formed. At first a low-pressure arc exists and very little light is produced; but gradually, as the lamp heats up, the mercury vapour pressure rises and a high pressure arc is formed and more light is emitted. Once the main arc is established, its resistance is lower than in the auxiliary electrode circuit and this auxiliary electrode ceases to function.

The time taken for the lamp to completely 'run up' – that is, to reach full light output – is approximately 5 min. Should the mains voltage be reduced to approximately 80 per cent or less of the normal value the lamp will be extinguished, and as the pressure is high it will not re-ignite unless a very high voltage is used or until the lamp has cooled and the pressure reduced. Although the subsequent run-up period is reduced, the total time to reach full light output again from switching off is of the order of 10 min.

MBF lamps have a fluorescent coating such as magnesium fluoro-germinate or yttrium vanadate on the inside of the outer envelope to convert the longwave UV radiation into visible energy, especially in the region of the spectrum above 600 nm, thus improving the spectral power distribution and, hence, the colour rendering.

Owing to the fairly good colour rendering, relatively high efficacy, and wide range of sizes available, this type of lamp is now extensively used in outdoor lighting and in industrial high bay installations. MBF lamps are also mixed with filament lamps in display lighting.

This type of lamp is also widely used for street lighting in residential areas where its colour rendering may be preferred to low-pressure sodium.

Lamp circuits

The circuit is relatively simple as the lamp incorporates its own starter. The circuit is shown in Fig. 4.19 and consists of a choke and power factor capacitor.

A series resistance in the form of a tungsten filament can be used instead of the choke. These are called MBT-type lamps; they have improved colour rendering but only about half the luminous efficacy. ‾ in common with all high-pressure lamps, cannot at present be satisfactorily di

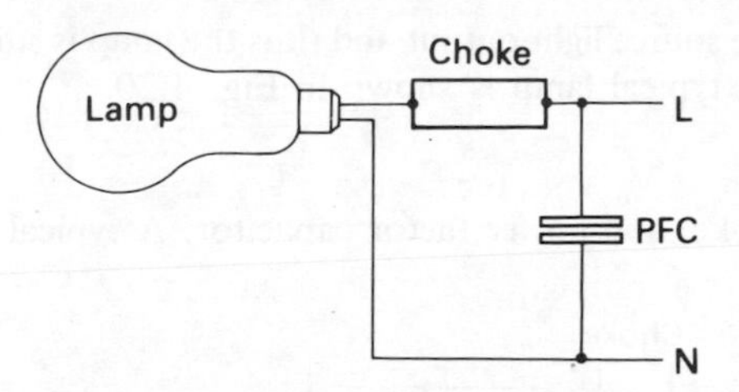

Fig. 4.19 Control circuit for an MBF lamp

Metal halide lamps (MBI)

This lamp, known as the mercury halide or metal halide lamp, is similar in construction to the MB lamp but has a much shorter arc tube. Strictly speaking, they are not mercury lamps but are a series of lamps combined within a single arc tube with output characteristics of different mixtures of elements. When such metals as dysprosium, thallium, indium, or sodium are introduced into the mercury arc tube, the mercury spectrum is suppressed (reducing the ultra-violet radiation), but the added metal atoms partly compensate by causing a wide range of spectral bands to be emitted. It is possible to tailor the spectral distribution to produce either a 'white' light or, alternatively, a predominantly coloured light (e.g. green), depending upon the relative proportions of the metals used. These metals have the disadvantage that, used in their elemental form, they would attack the discharge tube. Therefore, halides of the metal are used, the most common being the *iodide,* e.g. sodium iodide.

One disadvantage of these lamps is that a high starting voltage is necessary. This is usually provided by an electronic ignitor which generates very short duration high-voltage pulses and is connected across the lamp terminals. This type of starter enables the lamp to be restruck within one minute of being switched off.

Due to its high efficacy and good colour rendering properties, the MBI lamp is suited to both high bay interior and high mast exterior installations, and is often used in sports stadiums, especially where colour television transmission is expected. The small arc tube

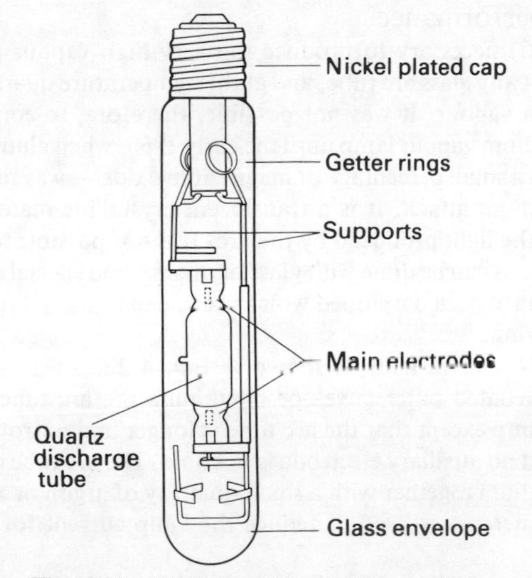

Fig. 4.20 375 W metal halide lamp (MBI)

enables good optical control of the source light output and thus the lamp is suitable for use in floodlighting projectors. A typical lamp is shown in Fig. 4.20.

Lamp circuits

The lamp requires an ignitor, choke, and power factor capacitor. A typical circuit is shown in Fig. 4.21.

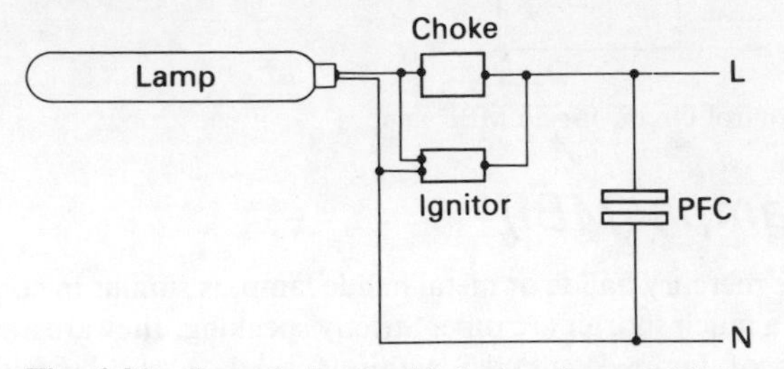

Fig. 4.21 Control circuit with ignitor for an MBI lamp

High-pressure sodium lamp (SON, SON-L, SON-R)

The low-pressure sodium arc emits all of its radiation in the two yellow lines at 589 and 589.6 nm. The colour rendering is, as stated earlier, very unsuitable for most applications other than road lighting and security lighting.

If the arc vapour pressure is increased beyond the optimum pressure, increased absorption of the 'D' lines occurs and the light output is reduced. However, as the pressure is increased and the 'D' lines disappear, other broad bands of energy appear in the spectrum until an almost continuous visible spectrum is emitted.

This type of source, because it produces radiant energy at the extremes of the visible spectrum where the V_λ value is low, has a lower efficacy of approximately 100 lm/W but provides much better colour rendering than the low-pressure lamp.

Note that the efficacy, despite the reduction by comparison with low-pressure sodium, is higher than can be obtained from mercury and metal halide sources.

Lamp construction and performance

The high temperature (1300 °C) necessary to produce the very high vapour pressure required in this lamp would melt any glass arc tube, and at this temperature quartz also is readily attacked by hot sodium vapour. It was not possible, therefore, to construct a practical SON high-pressure sodium vapour lamp until the early 1960s when **alumina** – a 'sintered' aluminium oxide with a small percentage of magnesium oxide – was first used.

Alumina can contain the sodium attack. It is a translucent crystalline material and transmits about 90 per cent of the light produced by the arc. It is not possible to pinch-seal the electrodes in the arc tube as can be done with glass or quartz, and special niobium caps containing the electrodes have been developed which are sealed to the arc tube ends by a sophisticated form of brazing.

A 400 W high-pressure SON sodium lamp is shown in Fig. 4.22.

The lamp consists of an evacuated outer envelope containing the arc tube and its support – similar to the MB lamp except that the arc tube is longer and narrower than the arc tube of the MB lamp and no auxiliary electrode is possible. The arc tube contains an amalgam of mercury and sodium together with a small quantity of argon or xenon to assist in striking the arc. The mercury is used to reduce the lamp current for a given wattage.

A high voltage is required across the lamp in order to strike the arc. This high voltage

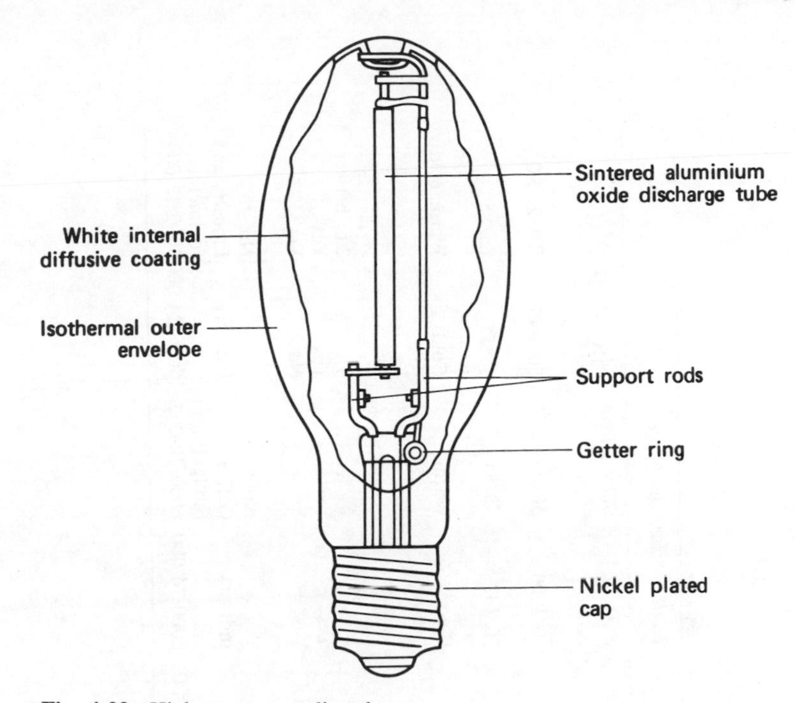

Fig. 4.22 High-pressure sodium lamp

may be provided, as in the MBI circuit, by an electronic ignitor which generates pulses of a few thousand volts for a few microseconds duration and is connected across the lamp terminals. The ignitor continues to pulse until the lamp is struck.

Some manufacturers include in the outer envelope of the lamp a bimetal starting switch and small heating element. The switch and element are in series and are connected across the electrodes.

This type of lamp is manufactured in numerous sizes from 125 W (76 lm/W) to 1 kW (130 lm/W) and the outer envelope may be either elliptical with a diffusing coating on the inside, or tubular clear glass.

The lamp sizes are similar to the high-pressure mercury vapour lamps and direct replacement may be made by using sodium lamps, providing the control gear is suitable. Some lamps are designed to be a direct replacement for a particular mercury vapour lamp without any major modifications to the circuit, e.g. a 330 W plug-in SON sodium lamp may replace a 400 W mercury lamp, giving approximately 25 per cent more light for 15 per cent less power (370 circuit watts against 420 watts). These replacement lamps, however, are not as efficient as the normal SON lamps and, therefore, are not recommended for new installations. The high-pressure sodium SON lamp is an economic light source and is used where colour rendering is not too critical. Typical applications are street lighting, outdoor area (car parks, etc.), and building floodlighting; also industrial lighting, especially in the high bay type installations. The increase in the cost of energy is causing an increasing number of indoor applications to be successfully and acceptably converted to high-pressure sodium sources.

Table 4.5 Summary of lamp types and their characteristics

	Tungsten filament		*Low-pressure discharge*		*High-pressure discharge*			
	GLS	*TH*	*MCF*	*SOX/SLI*	*SON*	*SON deluxe*	*MBF (HPL-N)*	*MBI (HPI)*
Wattage range (W)	25/2000	150/2000	4/125	10/180	50/1000	50/250	50/2000	70/2000
Efficacy range (lm/W)*	8/18	18/24	37/100	100/200	70/130	70/85	35/55	65/85
Colour appearance (CCT class)	Warm	Warm	Warm/ intermediate/ cold	Yellow	Warm/ yellow	Warm	Cold	Warm/cold
Colour rendering group	1A	1A	1A/3	—	4	2	3	2 or better
Need for ballast	No	No	Yes	Yes	Yes	Yes	Yes	Yes
Starter/ignitor	No	No	Yes†	Yes for SLI	Yes‡	Yes	No	Yes
Available with internal reflector	Yes	Yes	Yes	No	Yes	No	Yes	No
Typical use	Home	Display	Office and shop	Streets	Industry	Office (uplighting)	Industry	Industry and commerce
Average life (h)	1000/2000	2000/4000	5000/15 000	6000/20 000	6000/24 000	6000/12 000	5000/24 000	5000/12 000

* *This is lamp watts only.*
† *On switch-start circuits.*
‡ *Some lamps have internal switches.*

SON de luxe lamps

Further increase in the sodium vapour pressure causes a reduction in luminous efficacy but significant improvement in colour rendering. The deluxe lamps achieve an improvement in CRI from group 4 to group 2 and can be used in most interiors where a warm colour appearance is acceptable. The luminous efficacy is around 70 lm/W.

Lamp circuits

As with other discharge lamps the circuit comprises a starter, choke, and power factor capacitor. The starting process was outlined in the previous section.

Summary of lamps and their performance

No single table can show the complete range of lamps. Table 4.5 indicates the main groups of lamps, their range of availability, and their performance. Detailed information is only available from the manufacturers' catalogues.

5 Luminaires

The luminaire is the equipment which contains the lamp. Its purposes can be identified as:

(1) connecting the lamp to the electricity supply;
(2) controlling the light emitted by the lamp;
(3) protecting the lamp from a hostile environment;
(4) providing a fixture of satisfactory appearance.

The relative importance of each factor depends on the use. All luminaires must satisfy (1) and be electrically safe, but many decorative designs are far more concerned with (4) than (2).

Construction of luminaires

The shape of a luminaire and its method of construction are related to the lamp shape. Those for use with bulb-shaped lamps normally have a symmetrical form around the vertical axis, and their contours can be spun. This enables glass blowing and metal spinning to be used to considerable optical and decorative effect.

Tubular lamps need luminaires formed by bending, pressing, or extruding into linear shapes and their designs are usually in sheet steel or plastic.

Material and finishes

Steel

This can be fabricated by bending, pressing, or spinning. It is normally finished in stoved enamel which is based on the use of an alkyd resin. Where a very high standard of finish is needed an acrylic stoved enamel is generally used. Vitreous enamelling is the application of a thin layer of glass fused on to the metal surface at a temperature of about 800 °C. This gives good corrosion resistance but can only be applied to shapes that will not distort at high temperatures, and do not have sharp edges where the enamel will easily chip.

Aluminium

The light weight of this metal makes it suitable for spun shapes, but it is not rigid enough for general use in fluorescent luminaires. Its reflecting properties depend on its purity, the commercial grade having a reflectance around 0.7 and 0.8, and the super-purity grade having a reflectance of nearly 0.9. Although stoved enamel finish can be used, a good reflection is obtained by polishing the metal and anodising the surface to form a relatively hard oxide film, which may be transparent and which protects the reflector from corrosion. It is also used in cast form for exterior and industrial type luminaires. The grade of alloy is normally LM6 which does not need protective finishes.

Plastics

These can be divided into two broad groups, acrylics and polystyrenes.

Acrylic material is supplied in sheets or granules and is available in clear, opal, or coloured forms. It is shaped by pressing while softened by heat, or extruded into a continuous shape. It is a thermoplastic material.

Polystyrene is supplied in powder form and has similar properties but does not have such good stability and tends to be brittle. It is used in the cheaper ranges of diffusers and controllers.

Glass

Due to its weight, glass is little used in fluorescent luminaires. Its resistance to abrasion, its sparkle, and the fact that it can be blown in to many shapes and colours makes it in many respects preferable to plastics in decorative tungsten lighting. Glass is obtainable in both heat-resistant and heat-absorbent forms, and at present there seems little likelihood that glass will be replaced by plastic for vertical window glazing.

Optical design

Figure 5.1 shows the principal methods of optical control and Fig. 5.2 how these methods can be applied in the design of a typical luminaire for a fluorescent lamp.

Reflectors

Any surface which is not perfectly black will reflect light. The amount it reflects and the way in which it is reflected can be defined as the reflection property of that surface. Even a sheet of clear glass reflects some of the light incident on it (about 8 per cent). It is usually

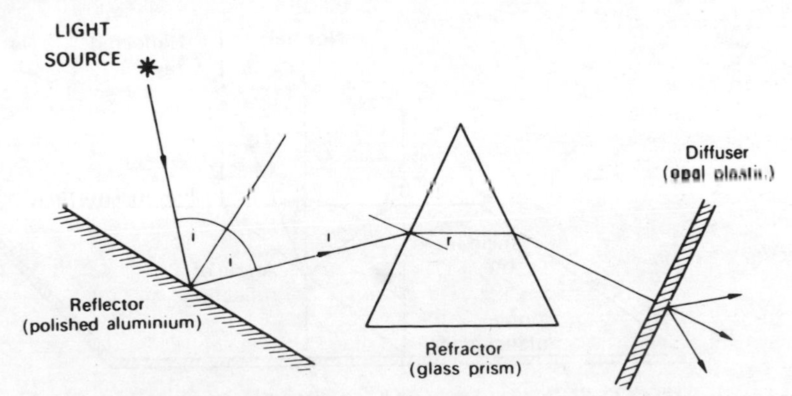

Fig. 5.1 Methods of light control

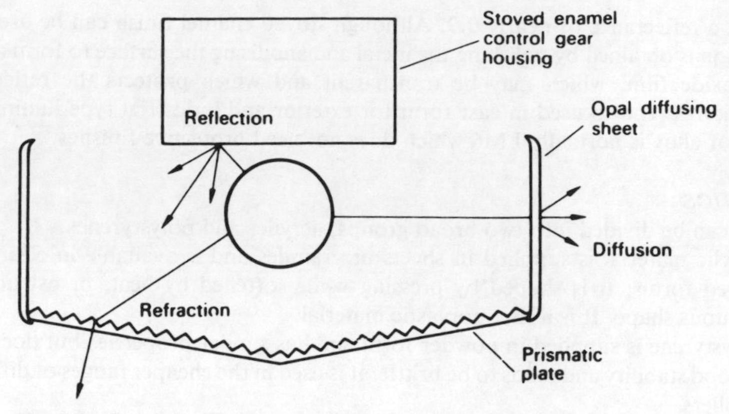

Fig. 5.2 Examples of light control in a fluorescent luminaire

the reflective properties of materials which reveal objects. The print is readable because the printing ink has different reflection properties from the paper on which it is printed. The light received is the same in both cases, but the light reflected differs.

Snell's law of reflection

The two laws of specular reflection (i.e. reflections from highly polished mirror types surfaces), are:

1. The *incident ray*, the *normal* to the surface, and the *reflected ray* are in the same plane.
2. The *angle of incidence* is the same as the *angle of reflectance.*

These are illustrated in Fig. 5.3 and they apply, whatever the shape of the reflector, at the point of incidence of the ray.

Specular reflectance

This is measured by the proportion of the light reflected in accordance with the laws of specular reflection. A good specular reflector, such as a silver-backed mirror, may have a specular reflectance near to 0.95, i.e. it *reflects* 95 per cent of the light it receives. Table 5.1 gives the reflectance of typical surfaces which can be used as specular reflectors.

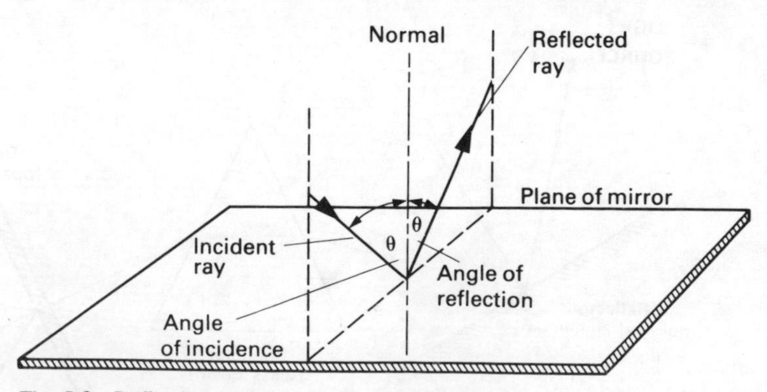

Fig. 5.3 Reflection of light on a specular surface

Table 5.1

Material	*Finish*	*Reflectance*
Aluminium (commercial grade)	Anodised and polished	0.7
Super purity	Anodised and polished	0.8
Aluminised plastic	Specular	0.94
Stainless steel	Polished	0.60
Chromium	Plate	0.66

Diffuse reflectance

If the surface is rough or non-shiny it will still reflect light (unless it is perfectly black), but this reflected light will obey the laws of specular reflection at minute irregular elements of the surface which, in effect, face different directions. It will be spread in various ways, and Fig. 5.4 illustrates different types of reflection that are identified.

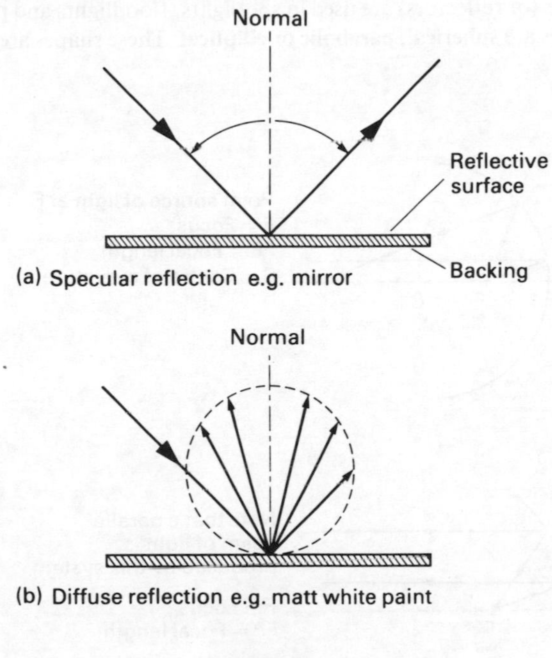

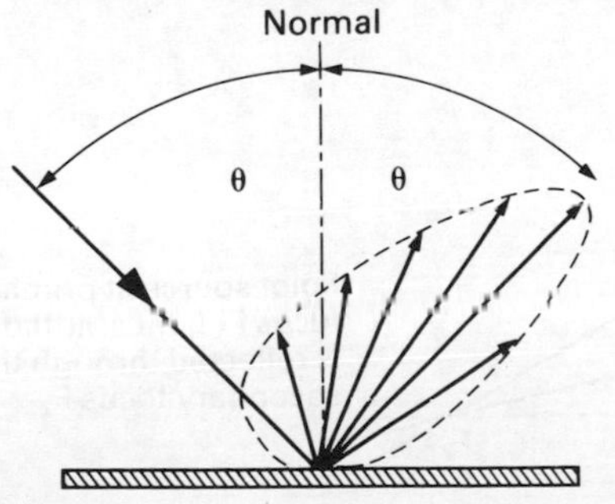

Fig. 5.4 Different types of reflection

Diffuse reflectance occurs when the light is scattered equally in *all* directions: a 'perfectly diffusing' reflector has the same luminance irrespective of the direction in which it is viewed. A surface which is almost a perfectly diffusing reflector is a block of magnesium carbonate. A more familiar surface is a sheet of blotting paper or chalk.

Mixed reflectance

Most surfaces have mixed reflection properties, i.e. some specular reflectance and some diffuse reflectance. A typical surface is the stove enamelling on the metal spine of a fluorescent luminaire.

Curved reflectors

Shaping the mirror can make the reflected light rays **converge** (concentrate) or **diverge** (spread). Curved mirrors (or reflectors) are used in spotlights, floodlights and projection systems. The basic curves are **spherical, parabolic** or **elliptical**. These shapes are shown in Fig. 5.5.

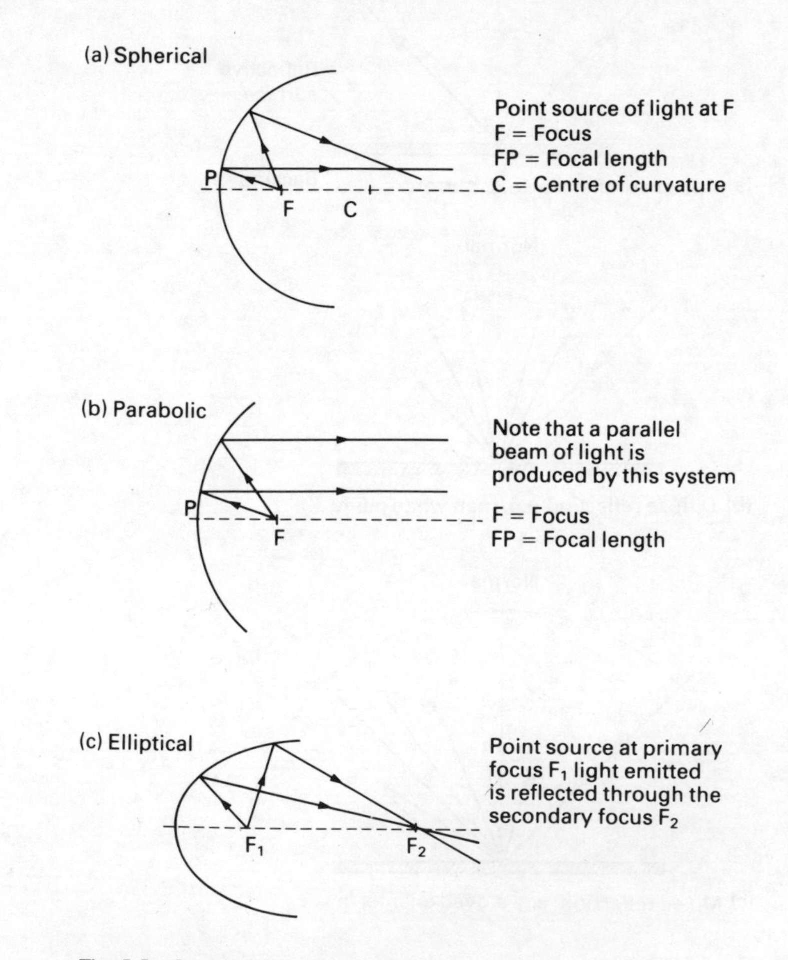

Fig. 5.5 Curved reflector contours

Spherical reflector

If a light source is placed at the focus F, which is at half of the radius from the centre, light reflected near P will emerge very roughly parallel.

If the light source were placed at its centre of curvature, C, then reflected light will come back through C. An example is the design of crown silvered lamp in spotlights used in display, as shown in Fig. 5.6.

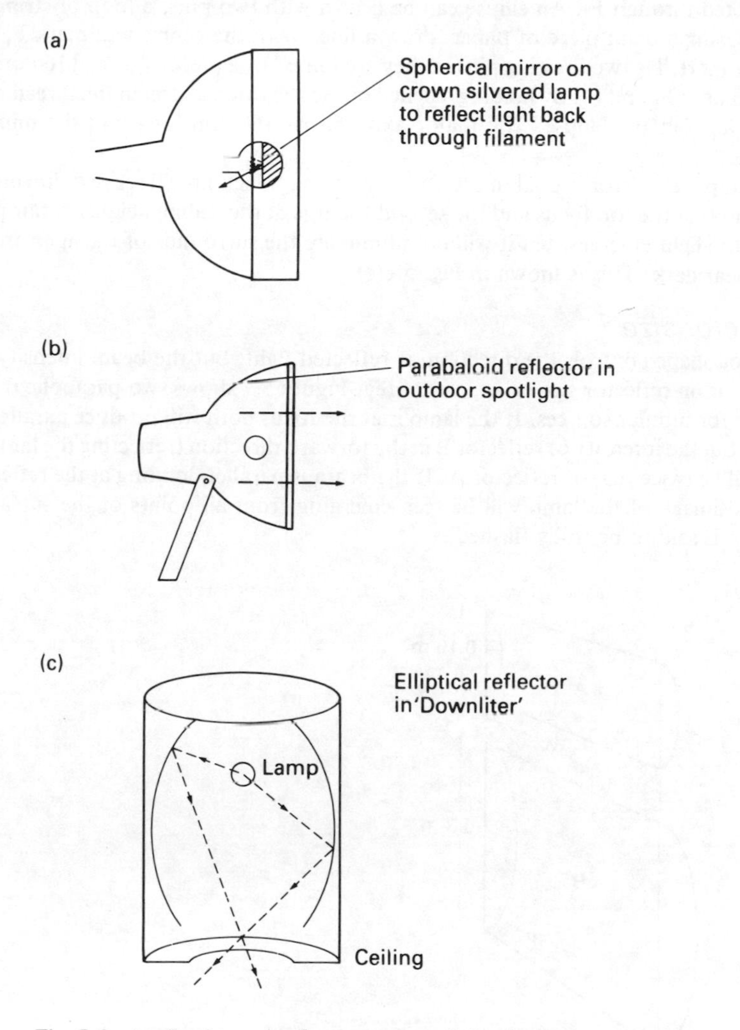

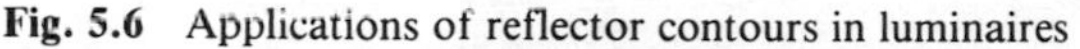

Fig. 5.6 Applications of reflector contours in luminaires

Parabolic reflector

The most important property of a parabola is that when a small light source is placed at its focus, all reflected light rays are parallel. If the source is moved away from the focus, the reflected rays will diverge and this will spread the beam of light. Parabolic reflectors are used for most spotlights and floodlights.

If the light source is tubular – e.g. a linear tungsten–halogen lamp – then the parabola is formed only in the plane at right angles to the lamp axis. The reflector is called a 'parabolic trough'. However, if the light source is comparatively small and of roughly equal dimensions in all directions – e.g. the filament of a GLS or projector lamp – then the reflector is parabolic in all planes, i.e. it is a 'paraboloid'.

Elliptical mirror

An ellipse has two focal points, F_1 and F_2, and if the light source is at F_1 all the light will be reflected through F_2. An ellipse can be drawn with two pins, a loop of string and a pencil. Using a plain piece of paper, draw a line down the centre and mark F_1 and F_2 100 mm apart. Tie two drawing pins so they are joined by a piece of thread 160 mm long. Fix the pins at F_1 and F_2. By holding a pencil inside the thread, stretch the thread out and hence trace out the ellipse. The major axis will be 160 mm long and the minor axis 120 mm.

The elliptical mirror is used in the design of some 'black hole' recessed downlighters. The lamp is at the top focus and the second focus is at the ceiling height. A fair proportion of the light emerges, but it will not illuminate the surrounds of the aperture which will appear dark. This is shown in Fig. 5.6(c).

Reflector size

Reflector shape controls the direction of reflected light, but the beam intensity is very dependent on reflector size or apparent area. Figure 5.7 shows two parabolic reflectors suitable for tubular sources. If the lamp is at the focus both will produce parallel beams of light but the intensity of reflector B in the forward direction (reflecting the lamp intensity) will be twice that of reflector A. If the beam is parallel, looking at the reflector, an enlarged image of the lamp will be seen emerging from all points of the surface. The reflector is said to be 'fully flashed'.

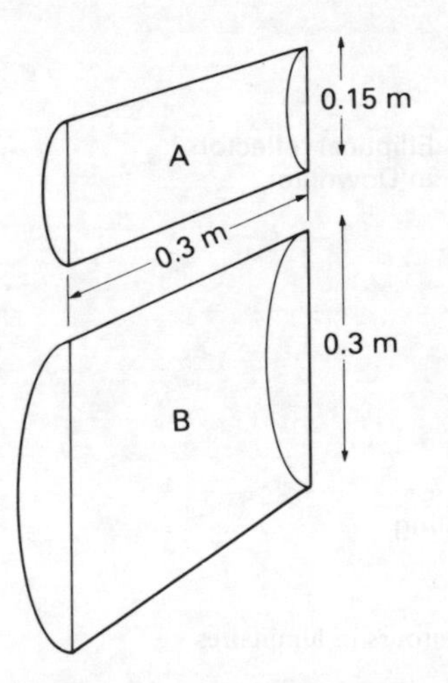

Fig. 5.7 Example of reflector size and intensity

Example
If the lamp in Fig. 5.7 has a luminance in all directions of 10 000 cd/m^2 transverse to its axis, and if the specular reflectance is 0.8, calculate the intensity for A and B.

Solution
The luminance of the image in the reflector in the forward direction will be $10\,000 \times 0.8$ cd/m^2.

The *intensity* in the forward direction will be:
Luminance times the *'projected' area* of the reflector

If the area of rectangle A is 0.3 m wide × 0.15 m high, and the area of rectangle B is 0.3 m × 0.3 m, then

$$I_A = 10\,000 \times 0.8 \times (0.3 \times 0.15)\text{ cd} = 360\text{ cd}$$
$$I_B = 10\,000 \times 0.8 \times (0.3 \times 0.3)\text{ cd} = 720\text{ cd}$$

Hence, I_B has double the intensity of I_A.

Note that the intensity of the lamp in the direction of view itself has been ignored because its contribution would be negligible in this type of reflector.

Refraction

Figure 5.8 shows the path of a ray of light through a clear glass prism. When light passes from one clear medium to another – e.g. glass–air, air–water – some light is reflected at the surface and some is transmitted. If the light enters the boundary between the two media along the normal (i.e. at right angles to the surface) its directions will not change. If it enters at an oblique angle, then its direction will change.

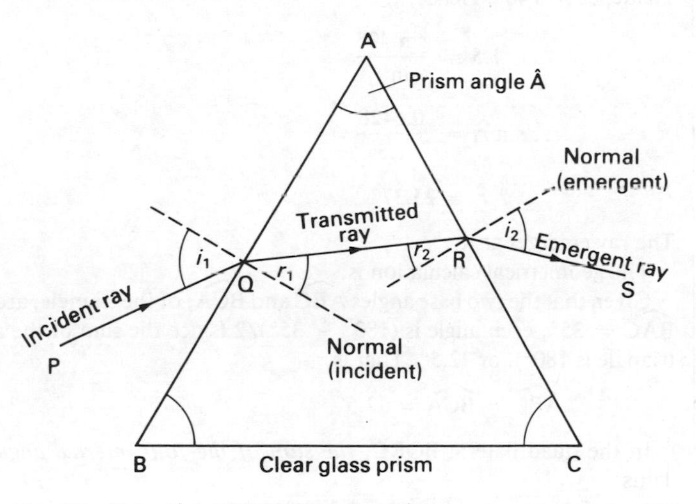

Fig. 5.8 Refraction of light through a prism

The angle of incidence ($\hat{i}$) is the angle the ray of light makes with the normal when passing from one clear material to another, e.g. from air to glass, plastic, or water. The angle of refraction ($\hat{r}$) is the angle the redirected ray makes with the normal.

When passing from a less dense material (e.g. air) to a denser material (e.g. glass) the angle of refraction ($\hat{r}$) is always less than the angle of incidence ($\hat{i}$).

$$\text{Refractive index } (\mu) = \frac{\sin \hat{i}}{\sin \hat{r}} \qquad \ldots [5.1]$$

Example
The refractive index (μ) of air to glass is 1.5. If the angle of incidence is 20°, what is the total deviation?

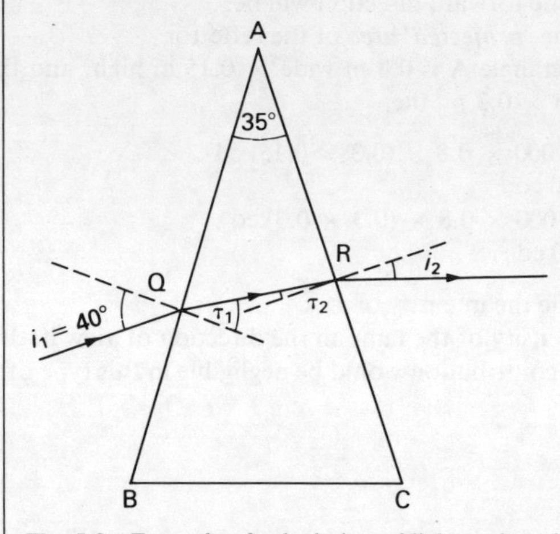

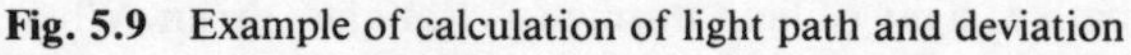
Fig. 5.9 Example of calculation of light path and deviation

Solution
Figure 5.9 shows a prism and the light path can be traced. At Q the angle of incidence $\hat{i}_1$ is 40°. Hence,

$$1.5 = \frac{\sin 40°}{\sin r_1}$$

$$\therefore \sin \hat{r}_1 = \frac{0.6428}{1.5}$$

$$\hat{r}_1 = 25.37°$$

The ray emerges at R.

The geometrical calculation is:

Given that the two base angles $\widehat{ABC}$ and $\widehat{BCA}$, of the triangle, are equal and $\widehat{BAC} = 35°$, each angle is $(180° - 35°)/2$ (since the sum of the angles of a triangle is 180°), or 72.5°. That is,

$$\widehat{ABC} = \widehat{BCA} = 72.5°$$

In the quadrilateral BQRC, *the sum of the four internal angles* is 360°. Thus

$$\widehat{BQR} = (90° + \hat{r}_1) = 90° + 25.37° = 115.37°$$

and

$$\widehat{QRC} = (360° - 115.37°) - (2 \times 72.5°)$$
$$= 99.63°$$

$$\therefore \quad \hat{r}_2 = 99.63° - 90°$$
$$= 9.63°$$

Again $$\frac{\sin \hat{i}_2}{\sin \hat{r}_2} = 1.5$$

$\therefore \quad \sin \hat{i}_2 = 1.5 \times \sin 9.63°$
$i_2 = 14.53°$

Now the ray of light would have been deviated by a total angle of:

$$(\hat{i}_1 + \hat{i}_2) - (\hat{r}_1 + \hat{r}_2) = (40° + 14.53°) - (25.37° + 9.63°)$$
$$= 19.53°$$

Total internal reflection occurs when the angle of incidence reaches a certain value. This means that instead of the light emerging at R it is totally reflected back into the prism.

This condition occurs when in theory the emergent angle i_2 closely approaches 90°. When this happens the glass–air barrier acts as a mirror. The angle (r) can be found since

$$\frac{\sin 90°}{\sin \hat{r}} = 1.5 = \mu \text{ (refractive index)} \qquad \ldots [5.2]$$

$\sin 90° = 1$, hence

$$\sin \hat{r} = \frac{1}{1.5} = 0.6667$$

$$\therefore \quad \hat{r} = 41.8°$$

This is known as the *critical angle* for a refractive index of 1.5. All transparent materials have their own critical angle. Prisms can therefore be made to *reflect* as well as *refract*.

Diffusion

Many luminaires make no use of reflection or refraction to control the light. A domestic lamp shade may provide the right sort of light distribution in an attractive way, but it is not the result of careful optical design!

It does alter the light distribution, partly by absorbing some of it, partly by reflecting some, and partly by scattering the light that it transmits. This process of scattering is called *diffusion of light*, and materials that do this, such as opal plastic, are called *diffusers*. These should not be confused with prismatic enclosures which, if correctly designed, should be controlling and not scattering the light.

Dichroic coatings

This is at present a rather specialised treatment of reflectors and lenses, but is has considerable significance in the filtering out of radiant heat. A dichroic mirror consists of a glass base on which alternate layers of transparent material are laid. Each layer has a different refractive index, e.g. magnesium fluoride (1.38) and zinc sulphide (2.30). If the optical thickness is $\lambda/4$, then interference will occur at that wavelength λ, and while the reflected light at each surface will be in phase, the transmitted light will be reduced to zero.

The thickness and refractive index of the layers dictate the values at which interference will occur, and so at certain values no interference will occur and most of the light will be transmitted.

A reflector or filter can be designed either to reflect or transmit certain colours, or it can reflect or transmit light but not infra-red. This provides a so-called 'cool beam', but it must be remembered that if the radiant heat is removed from the beam, it will emerge elsewhere, usually inside the luminaire.

Mechanical construction

Luminaires can take many different forms but all have to provide support, protection and electrical connection to the lamp. In addition, luminaires have to be safe during installation and operation and must be able to withstand the surrounding ambient conditions. The standard which covers most luminaires in the UK is BS 4533 *Luminaires*. It is suitable for use with luminaires containing tungsten filament, tubular fluorescent and other discharge lamps running on supply voltages not exceeding 1 kV. It covers the electrical, mechanical, and thermal aspects of safety. Luminaires should comply with BS 4533.

Luminaires are classified according to the degree of protection they provide against electric shock, the degree of protection against ingress of dust or moisture, and according to the material of the supporting surface for which the luminaire is designed.

Table 5.2 Classification of luminaires according to the type of protection provided against electric shock (from BS 4533); reproduced by permission.

Class	*Description*	*Symbol used to mark luminaires*
0*	A luminaire in which protection against electric shock relies upon basic insulation; this implies that there are no means for the connection of accessible conductive parts, if any, to the protective conductor in the fixed wiring of the installation, reliance in the event of a failure of the basic insulation being placed on the environment.	No symbol
I	A luminaire in which protection against electric shock does not rely on basic insulation only, but which includes an additional safety precaution in such a way that means are provided for the connection of accessible conductive parts to the protective (earthing) conductor in the fixed wiring of the installation in such a way that the accessible conductive parts cannot become live in the event of a failure of the basic insulation.	No symbol
II	A luminaire in which protection against electric shock does not rely on basic insulation only, but in which additional safety precautions, such as double insulation or reinforced insulation, are provided, there being no provision for protective earthing or reliance upon installation conditions.	
III	A luminaire in which protection against electric shock relies upon supply at safety extra low voltage (SELV) or in which voltages higher than SELV are not generated. The SELV is defined as a voltage which does not exceed 50 V a.c., r.m.s., between conductors or between any conductor and earth in a circuit which is isolated from the supply mains by such means as a safety isolating transformer or converter with separate windings.	III

** Class 0 luminaires are not permitted in the UK.*

Table 5.2 lists the luminaire classes according to the type of protection against electric shock. Class 0 luminaires are not permitted in the UK by reason of the Electrical Equipment (Safety) Regulations and the Electricity (Factories Act) Special Regulations 1908 and 1944.

The degree of protection the luminaire provides against the ingress of dust and moisture is classified according to the Ingress Protection (IP) System. This system describes a luminaire by a two-digit number in which the first digit classifies the degree of protection against the ingress of solid foreign bodies, from fingers to fine dust, and the second digit classifies the degree of protection against the ingress of moisture. Table 5.3 lists the classes of these two digits.

Table 5.4 lists the classification of luminaires according to the material of the supporting surface for which the luminaire is designed.

BS 4533 applies to most luminaires intended for use in neutral or hostile environments. It does not apply to many of the luminaires intended for use in hazardous environments, i.e. where there is a risk of fire or explosion. For such applications different standards and certification procedures apply. Detailed guidance on this topic can be found in the CIBSE Lighting Guide: *Hostile and hazardous environments.*

Table 5.3 The degrees of protection against the ingress of solid bodies (first characteristic numeral) and moisture (second characteristic numeral) in the Ingress Protection (IP) System of luminaire classification (from BS 4533; reproduced by permission)

First characteristic numeral	*Degree of protection*	
	Short description	*Brief details of objects which will be 'excluded' from the enclosure*
0	Non-protected	No special protection
1	Protected against solid objects greater than 50 mm	A large surface of the body, such as a hand (but no protection against deliberate access). Solid objects exceeding 50 mm in diameter
2	Protected against solid objects greater than 12 mm	Fingers or similar objects not exceeding 80 mm in length. Solid objects exceeding 12 mm in diameter
3	Protected against solid objects greater than 2.5 mm	Tools, wires, etc., of diameter or thickness greater than 2.5 mm. Solid objects exceeding 2.5 mm in diameter
4	Protected against solid objects greater than 1.0 mm	Wires or strips of thickness greater than 1.0 mm. Solid objects exceeding 1.0 mm in diameter
5	Dust protected	Ingress of dust is not totally prevented but dust does not enter in sufficient quantity to interfere with satisfactory operation of the equipment
6	Dust-tight	No ingress of dust

Degrees of protection indicated by the second characteristic numeral

Second characteristic numeral	Degree of protection Short description	Details of the type of protection provided by the enclosure
0	Non-protected	No special protection
1	Protected against dripping water	Dripping water (vertically falling drops) shall have no harmful effect
2	Protected against dripping water when tilted up to 15°	Vertically dripping water shall have no harmful effect when the enclosure is tilted at any angle up to 15° from its normal position
3	Protected against spraying water	Water falling as a spray at an angle up to 60° from the vertical shall have no harmful effect
4	Protected against splashing water	Water splashed against the enclosure from any direction shall have no harmful effect
5	Protected against water jets	Water projected by a nozzle against the enclosure from any direction shall have no harmful effect
6	Protected against heavy seas	Water from heavy seas or water projected in powerful jets shall not enter the enclosure in harmful quantities
7	Protected against the effects of immersion	Ingress of water in a harmful quantity shall not be possible when the enclosure is immersed in water under defined conditions of pressure and time
8	Protected against submersion	The equipment is suitable for continuous submersion in water under conditions which shall be specified by the manufacturer. *Note*: Normally, this will mean that the equipment is hermetically sealed. However, with certain types of equipment it can mean that water can enter but only in such a manner that it produces no harmful effects

Table 5.4 Classification of luminaires according to the material of the supporting surface for which the luminaire is designed (from BS 4533; reproduced by permission)

Description of class	*Symbol used to mark luminaires*
Luminaires suitable for direct mounting only on non-combustible surfaces	No symbol – but a warning notice is required
Luminaires without built-in ballast or transformer, suitable for direct mounting on normally flammable surfaces	No symbol
Luminaires with built-in ballast or transformer suitable for direct mounting on normally flammable surfaces	F

Luminaires in flammable atmospheres

Atmospheres where there is a risk of explosion are classified:

Zone 0 in which an explosive gas – air mixture is continuously present, or present for long periods.
Zone 1 in which an explosive gas – air mixture is likely to occur in normal operation.
Zone 2 in which an explosive gas – air mixture is not likely to occur in normal operation, and if it occurs it will exist only for a short time.

In Zone 0 there is no acceptable lighting system. Zone 1 also permits the use of flameproof equipment certified for use in a specific hazardous situation and 'increased safety protection' (e). Zone 2 permits a lesser degree of protection using non-spark luminaires (N). A brief description of each type is given below.

Pressurized systems (P)

These may be used in both Zone 1 and 2. In one type, air or an inert gas is maintained within the enclosure at a pressure sufficient to prevent ingress of the surrounding, possibly flammable, atmosphere to the enclosure. In another, a flow of air or of an inert gas is maintained to sweep away any flammable vapour that may enter the luminaire.

Both types require a system of interconnected enclosures with provision for switching off the electrical supply automatically in the event of failure to maintain the air pressure or flow.

A flameproof enclosure

This term is applied to luminaires certified by BASEEFA (British Approvals Service for Electrical Equipment in Flammable Atmospheres) and complying with BS 4683 Part 2 or the earlier standards BS 229 *Flameproof enclosures* and BS 889 *Flameproof electric lighting fittings*, at present under revision.

It is defined as

> '... able to withstand an explosion of the flammable gas or vapour which may enter it without suffering damage and without communicating the internal flammation to the external flammable gas or vapour for which it is designed through any joints or structural openings in the enclosure.'

It is necessary to control the length of path and gap width to cool the products of combustion and to prevent transmission from the inside to the outside surrounding atmosphere.

Increased safety or type of protection (e)

Until recently, only flameproof or pressurised luminaires were allowed in Zone 1 areas; now 'increased safety' type of protection (e) is permitted. This type of protection is covered by IEC recommendations. The protection is defined in BS 4683 Part 4 as follows:

> 'A method of protection by which additional measures are applied to electrical equipment so as to give increased security against the possibility of excessive temperatures and of the occurrence of arcs and sparks during the service life of the apparatus. It applies to electrical equipment, no part of which produce arcs, sparks or exceed the limiting temperatures in normal service'.

The light sources allowed are cold starting, fluorescent tubular lamps with single pin caps, tungsten filament lamps for general lighting service and mixed light (i.e. tungsten/mercury) lamps. The luminaire enclosure must also meet minimum strength requirements.

Type (e) luminaires may be used in Zone 1 and Zone 2 areas if the mechanical protection is at least equal to IP 54. The protection type (e) is suitable for use in all gases and vapours, as far as explosion risk is concerned, if the temperature class is acceptable and the materials of construction are compatible with vapours, etc., in the surrounding environment.

Non-sparking (N)

This class of luminaire is suitable for Zone 2 areas. Type of protection (N) is defined in BS 4683 Part 3 as

> '… a type of protection applied to electrical apparatus such that, in normal operation, it is not capable of igniting a surrounding explosive atmosphere and a fault capable of causing ignition is not likely to occur.'

It probably covers up to 80 per cent of hazardous areas likely to be encountered. An outline sketch of a typical class (N) luminaire is shown in Fig. 5.10.

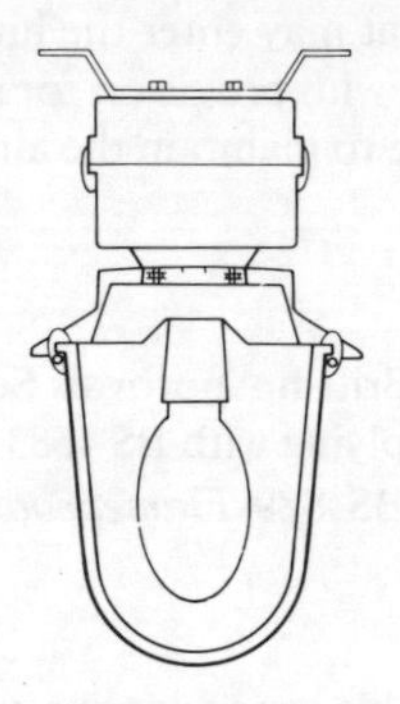

Fig. 5.10 Typical type 'N' luminaire for Zone 2 areas (courtesy Thorn EMI Lighting Ltd)

Photometric performance of luminaires

When selecting suitable luminaires for lighting schemes, many manufacturers' catalogues only provide a picture, some dimensions, lamp sizes, and perhaps the price. Other manufacturers may provide complete photometric data, and when carrying out any lighting design calculations it is sensible to base any predictions on luminaires of known performance. The requirements will be considered further in Chapter 7 but at this stage it is useful to consider the minimum information needed for lighting design calculations.

Light intensity distribution

This can be in polar diagram form. Figure 5.11 shows a typical fluorescent luminaire distribution. The left-hand curve is for the vertical plane through the lamp axis and the right-hand curve is for the vertical plane across the lamp axis (transverse plane). The candela scale should be in relation to a lamp output of 1000 lm. This value should then be adjusted for the output of the actual lamp (or lamps).

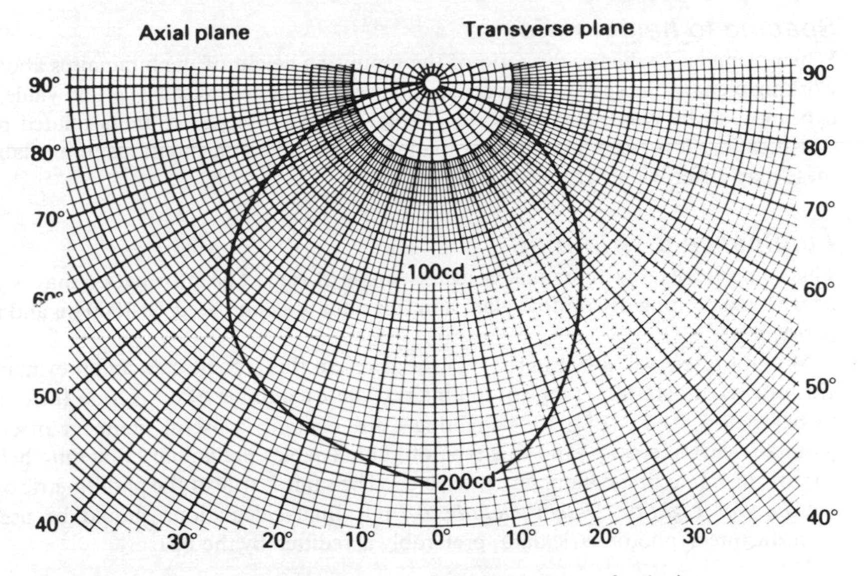

Fig. 5.11 Intensity distribution of a twin lamp reflector type luminaire

Example

The data in Fig. 5.11 applies to a twin lamp reflector type luminaire. What would be the intensity at 45° in the transverse plane if the lamps were two 65 W white fluorescent tubes?

Solution

The lumen output of the lamp can be found in the appendix (see p. 184). It is 4650 lm.

The transverse intensity at 45° = 100 cd.

$$\text{The correction factor} = \frac{2 \times 4650}{1000}$$

$$\text{Hence, the actual intensity} = 100 \times \frac{2 \times 4650}{1000} \text{ cd}$$

$$= 930 \text{ cd}$$

The intensity distribution can also be quoted in terms of the BZ classification, as explained in Chapter 2. This is sufficient for average calculations of illuminance in a room, but not for more precise calculations of illuminance at a specific point.

Light output ratios

These, as already explained in Chapter 2, are a measure of the light efficiency. All three values are normally quoted – i.e. DLOR, ULOR, and LOR. Combining light intensity distribution values and light output ratios it is fairly simple for a manufacturer to produce utilisation factor tables. This, again, is part of Chapter 7, but the data are available in CIBSE Technical Memorandum No. 5 to produce these tables.

Spacing to height ratios

When planning a scheme, the ratio of the mounting height of the luminaires above the working plane to the spacing between luminaires should not exceed a certain value if the lighting is to be reasonably uniform. Table 7.4 gives suggested limits related to BZ classification. It is essential to use the manufacturer's recommendations, if available, as these may differ widely from the table.

Luminance distribution

This comprises a table of luminance against angle of view. The information may be needed when calculating discomfort glare and this data is again based on 1000 lm and needs correction.

Manufacturers will differ in the way they present this type of data and they may provide one set of data for a range of luminaires where a number of correction factors have to be applied. The CIBSE 1984 Lighting Code contains a useful summary of typical luminaire characteristics, and this type of guide enables preliminary planning before a specific manufacturers' luminaire is selected. The same is true of the photometric data in the *Interior Lighting Design Handbook*, but final calculations should be used on manufacturers' photometric data, preferably acreditted by the BSI.

Production of photometric data Data sheets, as illustrated in Tables 7.3 and 7.9 of Chapter 7, are all derived from the intensity distributions of the luminaire, measured using an intensity distribution photometer. This is basically a photometric device that can rotate a photocell around the luminaire. It measures the illuminance which, using eqn [2.7], is converted to intensity. Most photometers are large because, for eqn [2.7] to be applied, the photometric distance should be five times the length of the source, e.g. for a 1.5 m luminaire the measuring distance is 7.5 m.

The method of photometry is set out in BS 5225 and the calculation of the data is set out in CIBSE Technical Memorandum No. 5 for utilisation factors and spacing-to-height ratios, and the CIBSE Technical Memorandum No. 10 — 'The Calculation of Glare Indices'.

Luminaire appearance

It is beyond the scope of the book to discuss the aesthetics of luminaire design, but the very nature of lighting is to attract attention to itself. The most efficiently designed lighting scheme can look dull, inappropriate, or quite offensive. The overall impression is a combination of the luminaire and the room appearance, and a lighting designer – be he or she an engineer, an architect, or an interior designer – should develop a sense of anticipating what the scheme will look like. Some people have a flair for this, but much can be achieved, or avoided, by looking critically at existing lighting and forming positive opinions about it.

6 Daylight

The vast majority of buildings contain windows, and for most of the working day the illuminance outside a building far exceeds that within. In considering the lighting of the interior a decision *should* be taken on the relative roles of daylight and electric light. In reality this seldom happens. The electric light will function adequately whether or not daylight is present.

However, as electricity costs have risen, this attitude has been challenged. Can daylight be effectively integrated with electric light? Methods have been devised, mainly, by the Building Research Establishment, producing in the 1960s Permanent Supplementary Artificial Lighting of Interiors (PSALI), a serious, if not particularly successful, attempt to produce integrated lighting design.

Work today is more concerned with economics and the savings that might be achievable. To achieve this it has to be possible to predict, with some degree of accuracy, the lighting levels that daylight will provide within a building throughout the year. This information is of interest both to the architect who designs the windows and to the engineer who designs the services.

Source of daylight

The source of daylight is the sun, emitting energy in a continuous spectrum similar to that of a black-body radiator at a temperature of approximately 6000 K.

This energy does not reach the earth's surface without change, as it is absorbed and scattered by the chemical elements, moisture and dust particles in the atmosphere. The scattering of the sun's energy by the atmosphere produces the blue sky which can then be considered as the effective source of daylight other than direct sunlight.

Luminance distribution of the sky

Before calculations of illuminance due to the sky can be made, the luminance of the sky must be determined. Originally it was assumed that the overcast sky was such a good diffusing medium that the luminance of all parts of the sky was the same; that is, that the sky had a uniform luminance. This type of sky is shown as **uniform sky** and under this condition the horizontal illuminance due to an unobstructed hemisphere of overcast sky is given by

$$E = L \times \pi \text{ lx} \qquad \ldots [6.1]$$

where L is the sky luminance in candelas per square metre, as shown in Fig. 6.1.

Measurements of sky luminance show that while the sky has a reasonably uniform luminance in azimuth, the luminance in any vertical plane varies. In 1955 the CIE adopted a non-uniform sky as defined by, $L_\theta = \frac{1}{3}L_z(1 + 2 \sin \theta)$ as standard, and this is known as the CIE Overcast Sky, where L_z is the luminance of zenith.

This form of sky is generally used for daylight calculations in a temperate climate, such as that in the UK.

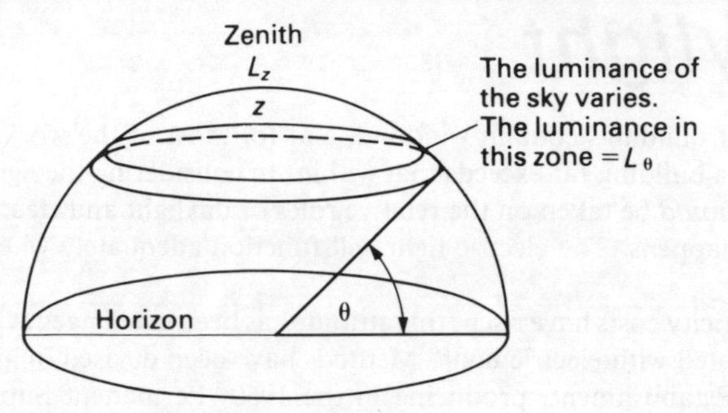

Fig. 6.1 The CIE overcast sky

Variability of daylight

The amount of direct sunlight is so variable and also so unpredictable in countries with climates like that of the UK that most of the calculations of illuminance due to daylight are limited to estimating the amount of light received from an overcast sky. Sunlight is, however, taken into account for estimation of solar heat gain, glare, and damage to works of art.

Variation in daylight illuminance

Measurements of illuminance due to the whole sky, excluding direct sunlight, have been made over a long period at various parts of the UK. Figure 6.2 shows the results of measurements made by the National Physical Laboratories. The maximum value of illuminance is about 35 000 lx and this occurs only in July and for only a relatively short period about noon. A standard outdoor illuminance of 5000 lx forms the basis for daylight calculations for the UK. Thus it is assumed that the average illuminance due to a complete hemisphere of sky – as measured, say, on the roof of a high building – is 5000 lx.

Daylight factor

An indication of the amount of daylight at a point within a room is the ratio of the daylight illuminance at the point to the instantaneous illuminance outside the building from a complete hemisphere of sky (excluding direct sunlight).

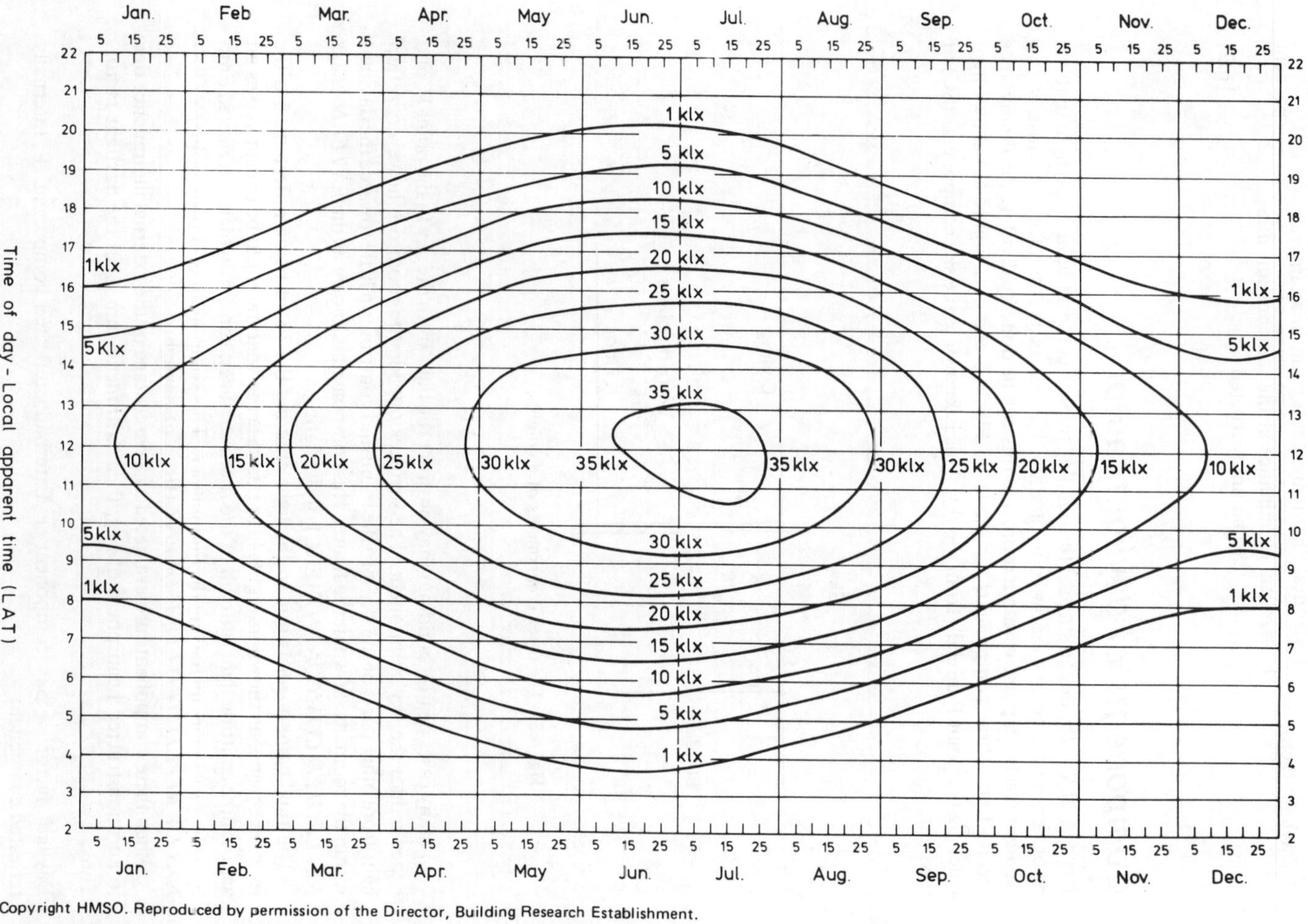

Copyright HMSO. Reproduced by permission of the Director, Building Research Establishment.

Fig. 6.2 Variation in illuminance (excluding direct sunlight) (Building Research Establishment. Crown copyright)

This ratio is known as the **daylight factor** (DF), and is usually quoted as a percentage:

$$\text{Daylight factor} = \frac{\text{Horizontal illuminance at a point in an interior}}{\text{Horizontal illuminance at the same instant due to an unobstructed sky}} \times \frac{100}{1} \text{ per cent} \quad \ldots [6.2]$$

Components of daylight factor

It is desirable at the design stage of the building to predict the amount of daylight that will be obtained for a given window configuration. It is necessary, therefore, to consider how daylight reaches a point within a room and this can be done by dividing the illuminance received into three components, as shown in Figs 6.3 and 6.4. Figure 6.3 shows light reaching the point P directly from the sky. This is known as the sky components (SC) of daylight.

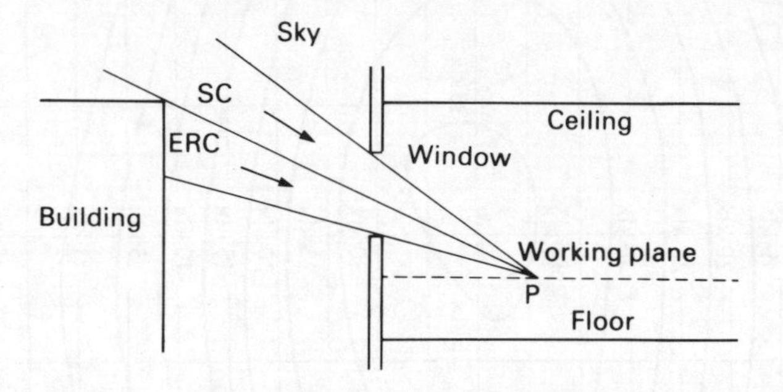

Fig. 6.3 The three components of daylight

In many cases nearby buildings may obstruct the light from the sky to the point P, and hence reduce the sky component. The surfaces outside the room do, however, reflect light from other parts of the sky into the room and contribute a little towards the daylight within the room, as shown in the figure. This component is known as the *EXTERNALLY REFLECTED COMPONENT* (ERC) of daylight.

The final component is due to the light entering the room being reflected onto the reference plane, as shown in Fig. 6.4. In this case the window could be considered as an area source emitting light onto all of the room surfaces, some of which is reflected onto the reference plane, increasing the illuminance. This component is known as the *INTERNALLY REFLECTED COMPONENT* (IRC) of daylight.

When these components are evaluated as a percentage of the external illuminance due to an unobstructed hemisphere of sky, their arithmetic sum gives the daylight factor. Thus,

Daylight factor = Sky component + Externally reflected component + Internally reflected component

or $$\text{DF} = \text{SC} + \text{ERC} + \text{IRC} \quad \ldots [6.3]$$

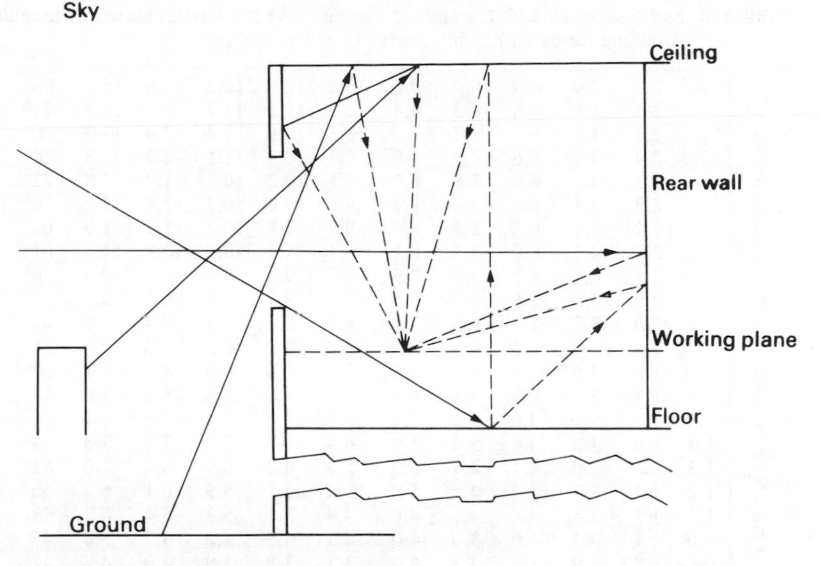

Fig. 6.4 The IRC component of daylight

Calculation of the components of daylight factor (DF)

The three components can be calculated by a range of methods. The SC and ERC are both found by considering the geometry of the visible sky or external reflected surfaces at a point on a horizontal plane within the room. IRC is based on inter-reflection theory and can be found from a formula, nomogram, or tables. The most commonly used methods in the UK are:

(a) *SC and ERC*: Waldram diagram; BRE protractors; BRE tables; and Pilkington 'pepper pot' diagrams.
(b) *IRC*: Formula; BRE tables; and BRE nomograms.

The presentation and explanation of these methods would require a separate book. This chapter refers only to tables and formulae. This is not because they are the easiest methods, but because they do not require special protractors or overlays. Computer programs are available for these calculations and would be the best choice if any number of calculations were necessary.

Calculation of sky component (SC)

Table 6.1 is used for determining these factors for windows with clear, vertical, rectangular glazing in conjunction with a CIE standard overcast sky. Other tables are available for other forms of glazing, e.g. roof lights.

The following information is needed:

1. H_1 and H_2, the heights of the window head and cill above the working plane.
2. W_1 and W_2, the distances of the window's vertical edges from a line drawn from the reference point for which the daylight factor is to be calculated, normal to the plane of the window.
3. D, the distance from the reference point to the plane of the window. (This is the plane of the inside of the wall, or the outside, whichever edge of the window aperture limits the view of the sky.)

Table 6.1 Sky components (CIE standard overcast sky) for vertical glazed rectangular windows (Building Research Establishment. Crown copyright)

Ratio H/D = Height of window head above working plane : Distance from window											Angle of obstruction
∞	2.5	4.9	6.9	8.4	9.6	10.7	11.6	12.2	13.0	15.0	90°
	2.4	4.8	6.8	8.3	9.4	10.5	11.1	11.7	12.7	14.2	79°
	2.4	4.7	6.7	8.2	9.2	10.3	10.9	11.4	12.4	13.7	76°
	2.4	4.6	6.6	8.0	9.0	10.1	10.6	11.1	12.2	13.3	74°
	2.3	4.5	6.4	7.8	8.7	9.8	10.2	10.7	11.7	12.7	72°
	2.3	4.5	6.3	7.6	8.6	9.6	10.0	10.5	11.4	12.3	70°
	2.2	4.4	6.2	7.5	8.4	9.3	9.8	10.2	11.1	11.9	69°
	2.2	4.3	6.0	7.3	8.1	9.1	9.5	10.0	10.7	11.5	67°
	2.1	4.1	5.8	7.0	7.9	8.7	9.1	9.6	10.2	10.9	66°
2.0	2.0	4.0	5.6	6.7	7.5	8.3	8.7	9.1	9.7	10.3	63°
	2.0	3.9	5.4	6.5	7.3	8.1	8.5	8.8	9.4	9.9	62°
	1.9	3.8	5.3	6.3	7.1	7.8	8.2	8.5	9.0	9.5	61°
	1.9	3.6	5.1	6.1	6.8	7.5	7.8	8.2	8.6	9.1	60°
	1.8	3.5	4.9	5.8	6.5	7.2	7.5	7.8	8.2	8.6	58°
1.5	1.7	3.3	4.6	5.6	6.2	6.8	7.1	7.4	7.8	8.1	56°
1.4	1.6	3.2	4.4	5.2	5.9	6.4	6.7	7.0	7.3	7.6	54°
1.3	1.5	2.9	4.1	4.9	5.5	5.9	6.2	6.4	6.7	7.0	52°
1.2	1.4	2.7	3.8	4.5	5.0	5.4	5.7	5.9	6.1	6.3	50°
1.1	1.3	2.5	3.4	4.1	4.6	4.9	5.1	5.3	5.4	5.7	48°
1.0	1.1	2.2	3.0	3.6	4.0	4.3	4.5	4.6	4.7	5.0	45°
0.9	0.99	1.9	2.6	3.1	3.4	3.7	3.8	3.9	4.0	4.2	42°
0.8	0.83	1.6	2.2	2.6	2.9	3.1	3.2	3.3	3.3	3.1	30°
0.7	0.68	1.3	1.7	2.1	2.3	2.5	2.5	2.6	2.6	2.8	35°
0.6	0.53	0.98	1.3	1.6	1.8	1.9	1.9	2.0	2.0	2.1	31°
0.5	0.39	0.70	0.97	1.10	1.3	1.4	1.4	1.4	1.5	1.5	27°
0.4	0.25	0.45	0.62	0.75	0.89	0.92	0.95	0.95	0.96	0.98	22°
0.3	0.14	0.26	0.34	0.42	0.47	0.49	0.50	0.50	0.51	0.53	17°
0.2	0.06	0.11	0.14	0.20	0.21	0.22	0.22	0.22	0.23	0.24	11°
0.1	0.02	0.03	0.04	0.05	0.05	0.06	0.06	0.06	0.07	0.08	8°
0	0.2	0.4	0.6	0.8	1.0	1.2	1.4	1.6	2.0	∞	0°
	Ratio W/D = Width of window to one side of normal : Distance from window										

Example

The SC is required at point A on Fig. 6.5.

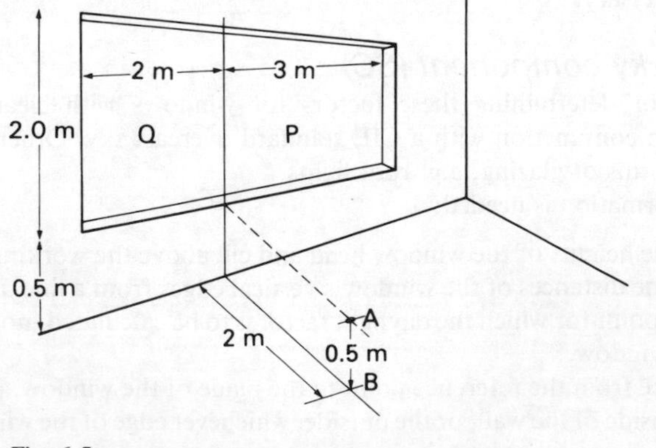

Fig. 6.5

Solution

Consider two separate window areas P and Q. In this example the plane of measurement is level with the window cill. D is 2 m, H_1 is 2.5 m, H_2 is 0.5 m, W_1 is 3 m, and W_2 is 2 m. In this case $H = H_1 - H_2$.

For window area P:

$$\frac{H}{D} = \frac{2.5 - 0.5}{2}$$

$$= 1$$

$$\frac{W}{D} = \frac{3}{2}$$

$$= 1.5$$

From Table 6.1 the SC is 4.55 per cent.

For window area Q:

$$\frac{H}{D} = \frac{2.5 - 0.5}{2}$$

$$= 1$$

$$\frac{W}{D} = \frac{2}{2}$$

$$= 1$$

Hence, from Table 6.1 the SC is 4 per cent.

∴ Total SC = 8.55 per cent.

The point A, however, may not be at the same height as the cill.

Example

The SC is required at floor level point B.

Solution

It is necessary to think of the window extending down to floor level, calculate the SC, then deduct the SC for the bit of 'extra' window between the cill and the floor.

For area P down to ground level:

$$\frac{H}{D} = \frac{2.5}{2}$$

$$= 1.25$$

$$\frac{W}{D} = 1.5 \text{ (as in previous example)}$$

For area Q down to ground level:

$$\frac{H}{D} = 1.25$$

$$\frac{W}{D} = 1 \text{ (as in previous example)}$$

Hence for this enlarged area:

$$\begin{aligned} SC &= 6.05 + 5.25 \\ &= 11.3 \text{ per cent} \end{aligned}$$

Now subtract the 'extra' window below P:

$$\frac{H}{D} = \frac{0.5}{2}$$

$$= 0.25$$

$$\frac{W}{D} = 1.5$$

and below Q:

$$\frac{H}{D} = 0.25$$

$$\frac{W}{D} = 1$$

$$\begin{aligned} \therefore \quad SC &= 11.3 - 0.7 \\ &= 10.6 \text{ per cent} \end{aligned}$$

It is interesting to note in these examples that although A is nearer the window than B, the SC is less. This is because the measurement is on a horizontal plane and at A very little light is seen or received from the lower part of the window. At B the window is more effective. Window height can be more significant than window width.

Calculation of externally reflected component (ERC)

If the direct entry of light through the window is severely limited by an external obstruction, it will be necessary to calculate the ERC. This can also be done using Table 6.1. The procedure is to treat the external obstruction visible from the reference point as a patch of sky whose luminance is some fraction of the sky obscured. In other words, the SC for the obstructed area is first calculated as described above and is then converted to the ERC by multiplying by the ratio of the luminance of the obstructed area to the sky luminance.

This is one of those calculations which is 'easier said than done'. The difficulty is to establish how much of the sky is obstructed, as seen from the point of measurement. Also, it is necessary to judge the reflectance of the obstruction. If it is 0.2 then the calculated ERC will be 0.2 of the equivalent SC.

Example

If in Fig. 6.5 the view from A shows a sky obstruction for the lower half of the window and the obstruction has a reflectance of 0.1, what percentage of sky and reflected sky reaches A?

Solution

The window must be considered as two sources. Taking the lower half, it effectively transmits light from a sky 0.1 the luminance of the real sky.

The SC for the lower half is calculated in the following way.

For window area P:

$$H = 1.5 - 0.5$$

$$\therefore \quad \frac{H}{D} = \frac{1.0}{2}$$

$$= 0.5$$

$$\frac{W}{D} = 1.5 \text{ (as before)}$$

For window area Q:

$$\frac{H}{D} = 0.5$$

$$\frac{W}{D} = 1 \text{ (as before)}$$

Hence SC = 1.4 + 1.6
= 3.0 per cent

This is for an open sky view, but as the view is obstructed,

ERC = SC × Reflectance
= 3.0 × 0.1
= 0.3 per cent

This is a small amount as might be expected in this situation, and illustrates how significantly external obstructions can reduce available daylight.

The SC will be for the top half of the window, but nearly all necessary calculations have already been done.

SC = SC for whole unobstructed window *minus* SC for lower half *if* unobstructed
= 8.55 − 3.0
= 5.55 per cent

If the window was totally obstructed

ERC = 8.55 × 0.1
= 0.8 per cent

∴ This makes the daylight of little value in terms of illumination.

'No-sky' line

If there are any significant external obstructions, a line can be drawn on the working plane in the room indicating the points at which direct sky view ceases. Such a line is shown on Fig. 6.6 which illustrates how the DF can be represented as a series of contours. When considering the lighting value of daylight it is suggested that the area beyond the 'no-sky' line will receive inadequate daylight for normal working illumination.

Calculation of internally reflected component (IRC)

If all the room surfaces were black, the daylight received at a point in the room would be that which comes direct from the window. In practice, all surfaces reflect light and this

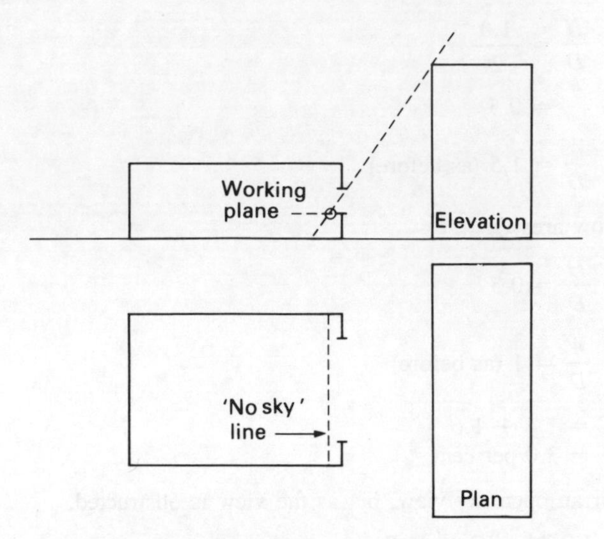

Fig. 6.6 The 'no-sky' line due to external obstructions

Table 6.2 The minimum IRC of DF (per cent) (Building Research Establishment. Crown copyright)

Ratio of window area to floor area	*Window area as percentage of floor area*	*Floor reflection factor*											
		10%				*20%*				*40%*			
		Wall reflection factor											
		20%	*40%*	*60%*	*80%*	*20%*	*40%*	*60%*	*80%*	*20%*	*40%*	*60%*	*80%*
1 : 50	2	—	—	0.1	0.2	—	0.1	0.1	0.2	—	0.1	0;2	0.2
1 : 20	5	0.1	0.1	0.2	0.4	0.1	0.2	0.3	0.5	0.1	0.2	0.4	0.6
1 : 14	7	0.1	0.2	0.3	0.5	0.1	0.2	0.4	0.6	0.2	0.3	0.6	0.8
1 : 10	10	0.1	0.2	0.4	0.7	0.2	0.3	0.6	0.9	0.3	0.5	0.8	1.2
1 : 6.7	15	0.2	0.4	0.6	1.0	0.2	0.5	0.8	1.3	0.4	0.7	1.1	1.7
1 : 5	20	0.2	0.5	0.8	1.4	0.3	0.6	1.1	1.7	0.5	0.9	1.5	2.3
1 : 4	25	0.3	0.6	1.0	1.7	0.4	0.8	1.3	2.0	0.6	1.1	1.8	2.8
1 : 3.3	30	0.3	0.7	1.2	2.0	0.5	0.9	1.5	2.4	0.8	1.3	2.1	3.3
1 : 2.9	35	0.4	0.8	1.4	2.3	0.5	1.0	1.8	2.8	0.9	1.5	2.4	3.8
1 : 2.5	40	0.5	0.9	1.6	2.6	0.6	1.2	2.0	3.1	1.0	1.7	2.7	4.2
1 : 2.2	45	0.5	1.0	1.8	2.9	0.7	1.3	2.2	3.4	1.2	1.9	3.0	4.6
1 : 2	50	0.6	1.1	1.9	3.1	0.8	1.4	2.3	3.7	1.3	2.1	3.2	4.9

**Assuming ceiling reflection factor = 70 per cent; angle of external obstruction = 20 degrees.*

reflected component can be significant. The calculation of the reflected component would be a tedious process but for the tables available.

Table 6.2 was formulated by the BRE to give minimum internally reflected components of daylight factor, assuming a CIE standard overcast sky and a known scheme of decoration. The table was designed primarily for rooms 6 m square (i.e. 36 m^2 in floor area) and 3 m ceiling height, having a window (glazed with ordinary glass) on one side extending from a 1 m cill to the ceiling, but by means of simple conversion factors, rooms of 19 – 100 m^2 floor area, with ceiling heights ranging from 2.5 to 4 m can be examined.

Table 6.3 Conversion factors to Table 6.2 for different flow areas and ceiling reflectances (Building Research Establishment. Crown copyright)

Floor area	*Wall reflection factor*			
	20%	40%	60%	80%
10 m²	0.6	0.7	0.8	0.9
100 m²	1.4	1.2	1.0	0.9

Initially, the ceiling reflection factor is assumed to be 70 per cent, but other values can be allowed for by means of conversion factors. A long external obstruction of 20° measured from the centre of the window is assumed as a common condition. In cases where there is more than one window in the room, the internally reflected components should be calculated separately for each window and then they are added together, remembering that the minimum internally reflected component for each window occurs at points in the room farthest from it.

Example
The room under examination is 6 m × 6 m (i.e. 36 m^2) × 3 m high and has a window in one wall with an area of 8.2 m^2. The floor has an average reflectance of 20 per cent; therefore, ratio of window area to floor area = 1 : 4.4.

The walls = 40 per cent reflectance
The ceiling = 70 per cent reflectance

Referring to Table 6.2 it will be seen that the minimum internally reflected component of DF lies between 0.6 and 0.8, or 0.7 per cent.

The conversion factor for ceiling reflectance is approximately 0.1 for 10 per cent change in reflectance below 70 per cent, e.g. for 50 per cent the reflection factor is 1.0 − 0.2 = 0.8.

The above method calculates the *minimum* value. It is usual practice to include this value in DF calculations even though the value of IRC will increase as the point of measurement moves closer to the window. When calculating the average DF, the average IRC can be found using the conversions in Table 6.4.

Table 6.4 Conversion factors for average internally reflected component (Building Research Establishment. Crown copyright)

Wall reflectance (%)	*Conversion factor*
20	1.8
40	1.4
60	1.3
80	1.2

Light losses

Allowance has to be made for light lost due to obstruction by window bars, transmission through the glass, and absorption due to dirt on the glass and room surfaces.

1. *Glazing bars*. Full window opening is assumed, and the reduction factors given in Table 6.5 should be applied.

Table 6.5 Light losses due to glazing bars (Building Research Establishment. Crown copyright)

For large-paned windows	*Reduction factor*
All metal windows	0.8
Metal windows in wood frames	0.75
All wood windows	0.65–0.70

2. *Glass transmission loss*. Allowance is made in the calculations for normal window losses. There are certain tinted glasses used for glare reduction and heat absorbing glasses for reduction in solar gain. A range of transmission factors is given in Table 6.6. Where double glazing is used, a factor of 0.9 should be applied.

Table 6.6 Relative transmittance and correction factors for various glazing materials (Building Research Establishment. Crown copyright)

Material	*Diffuse transmittance*	*Recommended correction factor to apply to daylight factor* *NOTE. Correction is proportional to transmission through clear glass*
Transparent glasses		
6 mm clear float	0.85	1.0
6 mm polished wired	0.80	0.95
Patterned and diffusing glasses		
3 mm rolled	0.80 to 0.85	0.95
6 mm rough cast	0.80 to 0.85	0.95
6 mm wired cast	0.75	0.90
Special glasses		
Heat-absorbing tinted float	0.45 to 0.75	0.50 to 0.90
Heat-absorbing tinted cast	0.50 to 0.65	0.60 to 0.75
Laminated insulating*	0.20 to 0.60	0.25 to 0.70
Plastics sheets		
Corrugated resin-bonded		
Glass-fibre-reinforced		
Roofing sheets		
moderately diffusing	* 0.75 to 0.80	0.90
heavily diffusing	* 0.66 to 0.75	0.75 to 0.90
very heavily diffusing	* 0.55 to 0.70	0.65 to 0.80
Diffusing plain opal 3 mm acrylic sheets	0.55 to 0.78 (depending on grade)	0.65 to 0.90

**Typical measurement values.*

3. *Maintenance factors*. Tables 6.7 and 6.8 give the factors as recommended in BRE Digest 42 to allow for dirt on the windows and on the room surfaces.

Applying these factors to the calculations:

Corrected DF = (SC + ERC) × Correction for glazing bars × Correction for type of glazing × Correction for dirt on glazing
+
(IRC) × Above corrections × Correction for dirt on room surfaces ... [6.4]

Table 6.7 Maintenance factors to be applied to the total DF or to each of its three components to allow for dirt on the glazing (Building Research Establishment. Crown copyright)

Location of building	*Inclination of glazing*	*Maintenance factor*	
		Non-industrial or clean industrial work	*Dirty industrial work*
Non-industrial or clean industrial area	Vertical	0.9	0.8
	Sloping	0.8	0.7
	Horizontal	0.7	0.6
Dirty industrial area	Vertical	0.8	0.7
	Sloping	0.7	0.6
	Horizontal	0.6	0.5

Table 6.8 Maintenance factors to be applied to the IRC of DF to allow for dirt or room surfaces (Building Research Establishment. Crown copyright)

Location of building	*Maintenance factor*	
	Non-industrial or clean industrial work	*Dirty industrial work*
Non-industrial or clean industrial area	0.9	0.7
Dirty industrial area	0.9	0.6

Example

Returning to the first example and assuming the IRC is 0.5 per cent, the window frames are metal, the windows are clear, double glazed, and the room is an office in an industrial area:

$$\text{Uncorrected DF} = \text{SC} + \text{ERC} + \text{IRC}$$
$$= 08.55 + 0 + 0.5$$
$$= 9.05 \text{ per cent}$$

Corrections:

Metal glazing = 0.8
Double glazing = 0.9
Dirt on windows = 0.8
Dirt in room = 0.8

$$\therefore \text{Corrected DF} = 8.55 \times 0.8 \times 0.9 \times 0.8 + 0.5 \times 0.8 \times 0.9 \times 0.8 \times 0.8$$
$$= 5.1 \text{ per cent}$$

Note that the corrections for a fairly normal situation have almost halved the DF.

Average daylight factor

For rooms with side windows the daylight falls rapidly as the distance from the window is increased. Figure 6.7 shows the fall off for a typical side lit room. Despite this variation, an average value can be calculated. This may appear to be meaningless, but work by

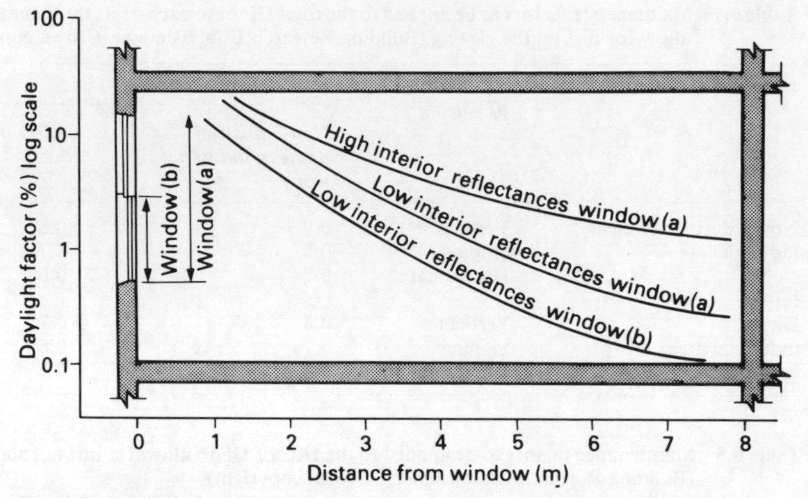

Fig. 6.7 Typical plots of DF against room depth for various window heights and reflectances

Lynes, Longmore, and others suggests that it can give an overall guide to the adequacy of the daylight.

The CIBSE 1984 Lighting Code suggests that when the average DF exceeds 5 per cent in a building which is used mainly during the day, electricity consumption for lighting should be too small to justify elaborate control systems on economic grounds, provided that switches are sensibly located. When the average DF is between 2 and 5 per cent, the electric lighting should be planned to take full advantage of available daylight. Localised or local lighting may be particularly advantageous, using daylight to provide the general lighting. When the average DF is below 2 per cent, supplementary electric lighting will be needed almost permanently.

Average DF will often give the designer sufficient information on which to base decisions on the relationship between natural and electric lighting. However, where more detailed information on the DF is necessary, the methods for point-by-point DF can be used.

A number of formulae are suggested for the calculation of the average DF. The simplest is known as the Littlefair/Plymouth expression, but is only suitable for broad assessments:

$$\text{Average DF} = \frac{\tau W \theta}{A(1 - R^2)} \qquad \ldots [6.5]$$

where

τ = diffuse transmittance of glazing material
W = net glazed area of window
A = area of all surfaces
R = average reflectance of all room surfaces
θ = the angle subtended, in the vertical plane normal to the window, by sky visible from the centre of the window. (This is shown in Fig. 6.8, and is quoted in degrees.)

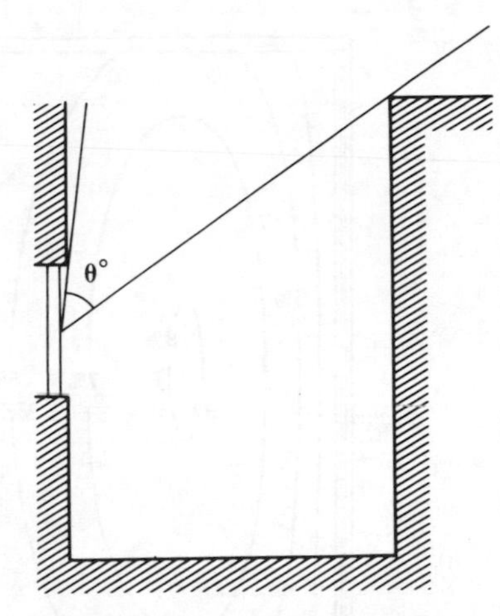

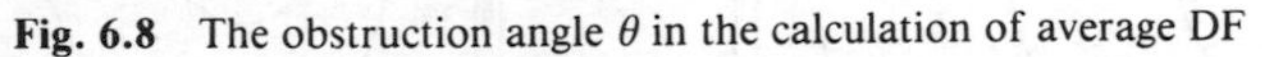

Fig. 6.8 The obstruction angle θ in the calculation of average DF

Orientation factors

The number of variables in daylight calculations can give rise to large errors. However, this is no excuse for over-simplification and approximation. One aspect under research is the error implicit in assuming a standard sky condition independent of building orientation. It is reasonable to assume that, even with overcast conditions, windows facing the direction of the sun will receive more light than those facing away. The CIBS 1984 Lighting Code contains a set of orientation factors (Table 6.9) which should be used when calculating the DF for energy-saving estimates, i.e.

$$\text{Orientation weighted DF} = \text{Orientation factor} \times \text{Overcast sky DF} \quad \ldots \ [6.6]$$

Table 6.9 Orientation factors for use with overcast sky daylight factors (Reproduced by permission of the Chartered Insititution of Building Services. Window design)

Direction	*Orientation factor*
South-facing window	1.20
East-facing window	1.04
West-facing window	1.00
North-facing window	0.77

Design of windows

By considering windows solely as admitters of daylight it may seem that the achievement of a certain minimum DF is the main design criterion. Electric lighting is today often used during daylight hours and it may be the case that in trying to achieve a minimum DF of, say, 2 per cent, the windows become too large causing more serious glare and heating problems. It may be preferable to design to a lower DF and integrate the daylight with the electric light.

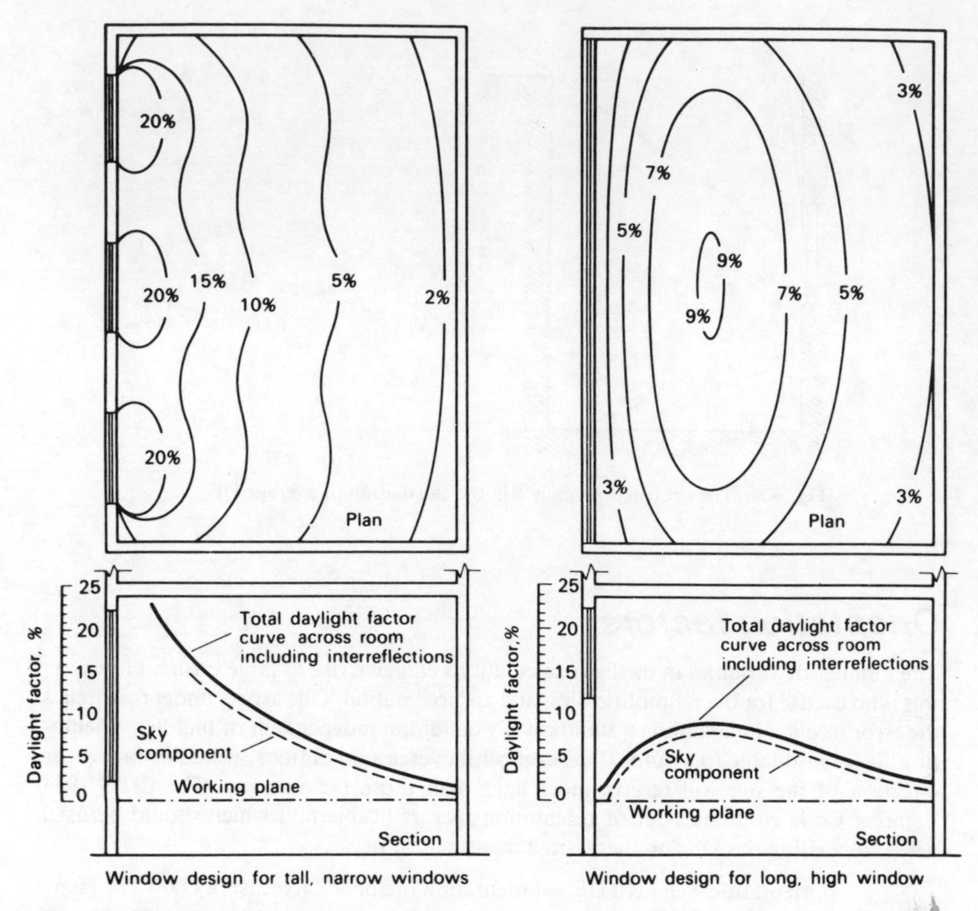

Fig. 6.9 The effect of window shape on DF

Figure 6.9 shows how the DF varies across a room using two types of windows, both having the same total area: (a) is for three tall windows, and (b) is for one long, high-level window. Window (a) will give a good open view, but (b) will provide more even daylight illumination and a higher minimum DF.

For rooflights, as shown in Figure 6.10, the daylighting is evenly spread over the working area and much higher factors are normally obtained. These factors are often drastically reduced by overhead obstructions and poor window maintenance.

To achieve an illuminance of 500 lx requires an average DF of around 10 per cent, which is much higher than could reasonably be expected in practice.

While DF is a useful guide to the light penetration from windows into rooms, it should not be the governing factor in window design if there is no economic justification for the large areas of glazing that might be involved.

J.A. Lynes has formulated a design process to establish a window design to achieve a required average DF. This is published in *Lighting Research and Technology* 11.102.79. This calculation is only relevant if the daylight contribution is significant, and to achieve this he suggests:

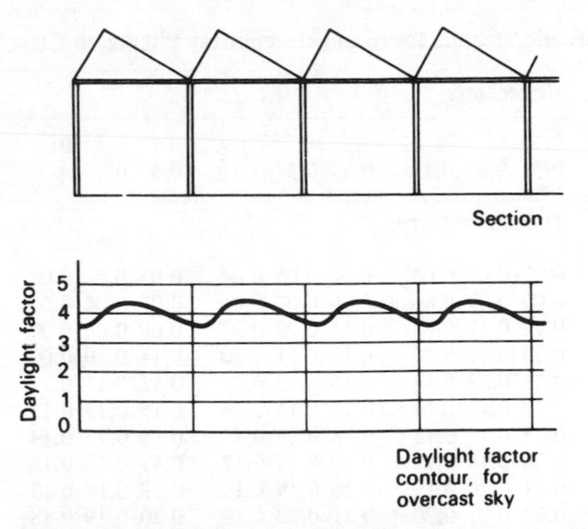

Fig. 6.10 The distribution of daylight across the floor for a roof-lit interior

1. No significant part of the working plane shall lie beyond the 'no-sky' line.
2. $(L/W + L/H)$ shall not exceed $(2/(1 - R_B))$, where L = depth of room from window to back wall; W = width of room, measured parallel to window; H = height of window head, above floor level; and R_B = average reflectance of the half of the interior remote from the window

If these conditions cannot be satisfied the daylighting will be unsatisfactory whatever the size of the window as the back of the room will always look gloomy compared with the space by the window.

Calculation of daylight factor for rooflights

It is normally assumed that if the light-transmitting structure of the roof is evenly spaced, then the daylighting at floor or working plane level will be reasonably uniform (see Fig. 6.10). A single calculation is then sufficient to determine the average DF.

A formula to find this value is:

$$\text{DF} = 100 \times \text{UF} \times M \times B \times G \times k \text{ per cent} \qquad \ldots [6.7]$$

where.

UF is the utilisation factor of the rooflights as given in Table 6.10 for various reflectances and room indices, and

$$\text{Room index} = \frac{\text{Length} \times \text{Width of room}}{(L + W) \times \text{Roof height above working plane}}$$

M is a light loss factor for dirt on the glass (Table 6.7)
B is a correction factor for glazing bars and internal obstructions (Table 6.5)
G is a correction factor for the type of glazing (Table 6.6)
k is the ratio of glazing area to floor area.

Example

A factory bay is 20 m × 10 m with a twin sloping monitored roof of average height 6 m above the floor. If the reflectances are 0.7 for the ceiling and 0.3 for the walls and k is 0.2, calculate the average DF at floor level. (The glass is 6 mm wired cast.)

Table 6.10 Example of utilisation factors for rooflights (courtesy Pilkington Glass Ltd)

	Reflectance								
Ceiling	0.7			0.5			0.3		0
Walls	0.5	0.3	0.1	0.5	0.3	0.1	0.3	0.1	0
Room index	*Utilisation factor*								
0.6	0.07	0.05	0.04	0.06	0.05	0.04	0.05	0.04	0.03
0.8	0.09	0.07	0.06	0.09	0.07	0.06	0.07	0.06	0.05
1.0	0.12	0.10	0.08	0.11	0.09	0.08	0.09	0.08	0.07
1.25	0.14	0.12	0.10	0.13	0.11	0.10	0.11	0.10	0.09
1.5	0.15	0.13	0.12	0.15	0.13	0.12	0.13	0.11	0.11
2.0	0.17	0.15	0.14	0.16	0.15	0.14	0.15	0.13	0.13
2.5	0.18	0.17	0.15	0.18	0.16	0.15	0.16	0.15	0.14
3.0	0.20	0.18	0.17	0.19	0.18	0.17	0.17	0.16	0.16
4.0	0.21	0.20	0.19	0.20	0.19	0.19	0.19	0.18	0.17
5.0	0.22	0.21	0.20	0.21	0.20	0.19	0.20	0.19	0.18
Infinity	0.25	0.25	0.25	0.25	0.25	0.25	0.24	0.24	0.23

Twin vertical monitor

Solution

Some assumptions will have to be made concerning the location and obstructions.

To find UF:

$$\text{Room index} = \frac{10 \times 20}{(10 + 20) \times 6}$$

$$= 1.1$$

So from Table 6.10 the reflectances 0.7 and 0.3 and a twin sloping monitor roof:

$$\text{UF} = 0.11$$

From Table 6.7 for sloping glazing and clean industrial work and area:

$$M = 0.8$$

From Table 6.5 for metal glazing bars and assuming no internal obstructions:

$$B = 0.8$$

From Table 6.6 for 6 mm wired cast glass:

$$G = 0.9$$

k is given as 0.2.

$$\therefore \quad \text{DF} = 100 \times 0.11 \times 0.8 \times 0.8 \times 0.9 \times 0.2$$

$$= 1.3\%$$

Sunlight in buildings

All calculations have referred to an overcast sky and have excluded direct sunlight. It is not feasible to consider direct sunlight as a positive factor when designing interior lighting. Much as it is welcomed in home interiors, considerable effort is taken to exclude it from commercial and industrial buildings.

The exclusion can take various forms including blinds, solar reflecting or absorbing glasses, or louvres. All these will modify the DF and often nullify the lighting value of the glazing.

Any detailed discussion on sunlight calculations is outside the scope of this book. The geometry of sunlight penetration is explained in such documents and books as BS DD67: 1980 *Basic data for the design of buildings: Sunlight* and *Windows and the Environment* (Pilkington Advisory Service).

CIBSE Window Design Guide 1987

This guide attempts to coordinate existing thinking on windows and their role in the building environment, visually, thermally, and acoustically. The main text deals with the process of window design as a transmitter of daylight into buildings. The appendix, which is longer than the design section, brings to one publication current data on prediction of daylight and sunlight and other environment topics.

7 Design of general lighting schemes

This should be the culmination of all the knowledge gained from the previous chapters. Unfortunately, this is where many people responsible for the design of lighting start and finish. Good design is the application of the best or most appropriate equipment in an economical but effective manner.

This chapter covers the normal design processes for establishing a satisfactory illuminance on the working plane, provided in a glare controlled manner. The overall pattern of room illuminances is considered in relation to suggestions contained in the CIBSE 1984 Lighting Code. Finally, the requirements for emergency lighting are outlined. The chapter contains a limited amount of photometric data, sufficient to indicate how it is used, but the larger manufacturers will usually provide full data for their range of luminaires.

How much light?

Standard values of illuminance (lux) are specified in the CIBSE 1984 Lighting Code for all the usual locations and occupations. These values are necessary to achieve:

(1) satisfactory illuminance of the task;
(2) an agreeable general appearance of the interior.

They are quoted as 'standard service illuminances', normally on a horizontal working plane. A 'service' illuminance is the mean illuminance throughout the life of the lighting system and averaged over the relevant area. 'Standard' implies a value suitable for many locations, unless there are special requirements. Table 7.1 shows the illuminances relevant to a range of different visual tasks.

This table is expanded in the Guide to quote recommended standard service illuminance values covering nearly 150 activities or types of interiors.

The illuminance values serve as a guide to good practice. They are not mandatory although some authorities may adopt them as part of their standard requirements. It is also recognised that there can be circumstances for a given situation where the standard service value should be modified to give a design service value, i.e. one applicable to the specific circumstances.

This is demonstrated in the form of a flow chart, and is reproduced in Table 7.2.

Table 7.1 Examples of activities/interiors appropriate for each standard service illuminance (reproduced by permission of the Chartered Institution of Building Services Engineers. CIBSE 1984 Interior Lighting Code).

Standard Service Illuminance (lx)	*Characteristics of the activity/interior*	*Representative activities/interiors*
50	Interiors visited rarely with visual tasks confined to movement and casual seeing without perception of detail.	Cable tunnels, indoor storage tanks, walkways.
100	Interiors visited occasionally with visual tasks confined to movement and casual seeing calling for only limited perception of detail.	Corridors, changing rooms, bulk stores.
150	Interiors visited occasionally with visual tasks requiring some perception of detail or involving some risk to people, plant or product.	Loading bays, medical stores, switchrooms.
200	Continuously occupied interiors, visual tasks not requiring any perception or detail.	Monitoring automatic processes in manufacture, casting concrete, turbine halls.
300	Continuously occupied interiors, visual tasks moderately easy, i.e. large details > 10 min arc and/or high contrast.	Packing goods, rough core making in foundries, rough sawing.
500	Visual tasks moderately difficult, i.e. details to be seen are of moderate size (5–10 min arc) and may be of low contrast. Also colour judgement may be required.	General offices, engine assembly, painting and spraying.
750	Visual tasks difficult, i.e. details to be seen are small (3–5 min arc) and of low contrast, also good colour judgements may be required.	Drawing offices, ceramic decoration, meat inspection.
1000	Visual tasks very difficult, i.e. details to be seen are very small (2–3 min arc) and can be of very low contrast. Also accurate colour judgements may be required.	Electronic component assembly, gauge and tool rooms, retouching paintwork.
1500	Visual tasks extremely difficult, i.e. details to be seen extremely small (1–2 min arc) and of low contrast. Visual aids may be of advantage.	Inspection of graphic reproduction, hand tailoring, fine die sinking.
2000	Visual tasks exceptionally difficult, i.e. details to be seen exceptionally small (< 1 min arc) with very low contrasts. Visual aids will be of advantage.	Assembly of minute mechanisms, finished fabric inspection.

Designing a general layout

The first stage is to decide on the type of luminaire to be used. This decision may seem premature, but, unless using computer programs, it is difficult to achieve anything without immediate access to specific photometric data. Alternative approaches can be

Table 7.2 Flow chart for obtaining the design service illuminance from the standard service illuminance (reproduced by permission of the Chartered Institution of Building Services Engineers. CIBSE 1984 Interior Lighting Code).

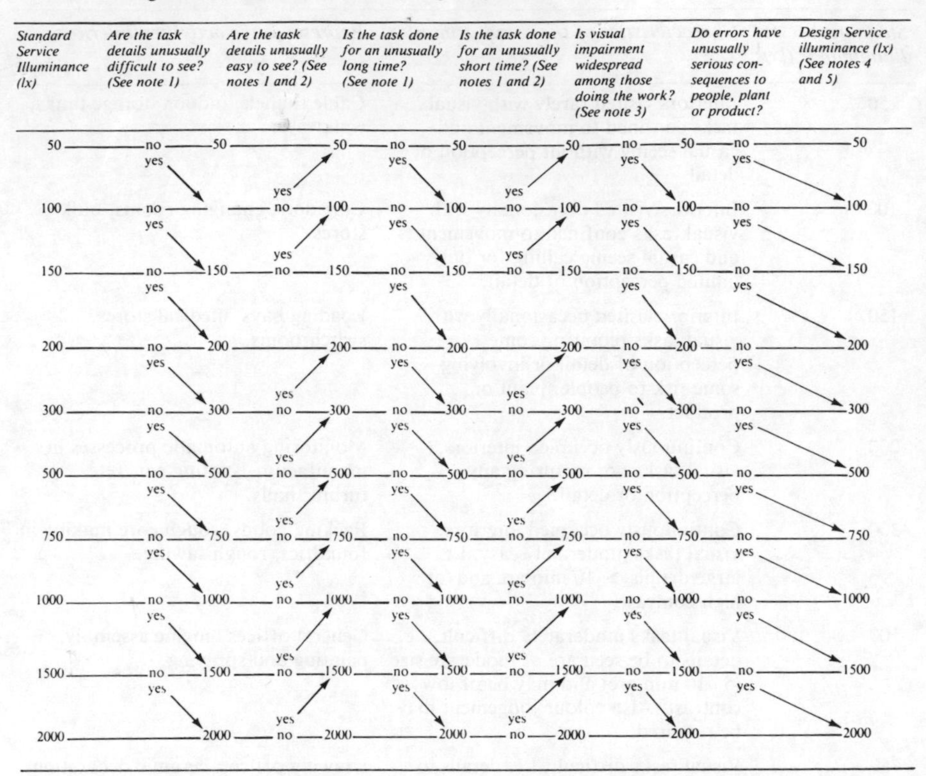

Notes

1. *The standard service illuminances recommended in the schedule are based on tasks which are representative of their type in the detail that has to be seen and the time for which the task has to be done. These steps in the flow chart allow for departures from these assumed conditions.*
2. *The standard service illuminance of 200 lx is provided as an amenity for continuously occupied interiors, even when perception of task detail is not required.*
3. *If the cause of visual impairment is dirty or scratched spectacles, safety glasses, safety screens, etc., it may be more effective to clean or replace these items rather than change the lighting. If safety screens are acting as a source of veiling reflections then the lighting/task/worker geometry should be re-arranged.*
4. *If the design service illuminance is more than two steps on the illuminance scale above the standard service illuminance, consideration should be given to whether changes in the task details, the organisation of the work or the people doing the work are more appropriate than changing the lighting.*
5. *For a design service illuminance of 1500 lx or 2000 lx, local lighting supplemented by optical aids should be considered.*

used where all the design requirements are established, and luminaires are then found to fit these requirements. This method is given in CIBSE Technical Report No. 15: *Multiple criterion design*. However, it is easier to explain the design process by the first method.

Next, the total number of lamp lumens is calculated to achieve the recommended illuminance, and finally the layout is planned.

Determination of lumens required

A commonly accepted method involves the use of the 'lumen formula' which is:

$$\text{Lamp lumens required} = \frac{E \times A}{\text{UF} \times \text{LLF}} \quad \ldots [7.1]$$

where E = standard service illuminance (lx)
A = area of working plane (m^2)
UF = utilisation factor
LLF = light loss factor

Utilisation factor

This is a measure of the efficiency of the lighting scheme and is the proportion of lamp flux that reaches the working plane. Some will come direct and some will come after reflection from other room surfaces.

Table 7.3 shows a typical set of data, and to determine the precise value involves details of the room dimensions and surface reflectances.

The room dimensions are required in terms of the room index, where

$$\text{Room index} = \frac{\text{Length} \times \text{Width}}{(\text{Length} + \text{Width}) \times \text{Height of lamp above working plane}} \quad \ldots [7.2]$$

Light loss factor and maintenance factor

In previous lighting codes the effect of hours of operation on the lumen output from lamps has been allowed for by using lighting design lumens rather than initial lumens, and the effect of dirt depreciation has been allowed for by using a maintenance factor. The maintenance factor is the ratio of the illuminance provided when the installation is in an average condition of dirtiness to the illuminance provided when it is clean.

This system has been abandoned because with the range of light sources now available it has become unrealistic to consider the illuminance provided at 2000 h as an accurate description of the average through-life illuminance provided by the installation. Dealing with maintenance by means of the light loss factor allows for a more comprehensive examination of the effect of the operating conditions and maintenance procedures on the illuminance provided.

Determination of light loss factor

Light loss factor is the product of three other factors:

$$\begin{aligned}\text{Light loss factor} = {} & \text{Lamp lumen maintenance factor} \\ & \times \text{Luminaire maintenance factor} \\ & \times \text{Room surface maintenance factor} \quad \ldots [7.3]\end{aligned}$$

Lamp lumen maintenance factor estimates the decline in light output of the light source over a specified time. Luminaire maintenance factor estimates the effect of dirt deposited on or in the luminaire over a set time on the light output of the luminaire. Room surface maintenance factor estimates the effect of dirt deposited on the room surfaces over a set time on the illuminance produced by the installation.

By calculating the light loss factor for different times, and taking into account the proposed maintenance schedule, it is possible to predict the pattern of illuminance produced by the installation over time. This pattern can be used to assess the suitability of any proposed maintenance schedule. It can also be used to estimate the average

Table 7.3 Photometric data for Thorn EMI prismatic enclosure type luminaire to be used with a Range of fluorescent lamps

Description	Single prismatic controller 1800 mm
Report number	500/IL/5295/1
Light Output Ratio	Up .26 Down .53 Total .79
Max. Spacing to Height Ratio (SHR MAX)	1.92

Luminous intensity (cd/1000 lm)			*Aspect factors*		
Angle (degrees)	*Transverse plane (T)*	*Axial plane (A)*	*Angle (degrees)*	*Parallel plane*	*Perpendicular plane*
0	132	132	0	0.000	0.000
5	132	131	5	0.087	0.004
10	139	130	10	0.173	0.015
15	147	126	15	0.256	0.033
20	155	122	20	0.334	0.058
25	160	116	25	0.407	0.088
30	160	109	30	0.473	0.123
35	155	102	35	0.532	0.160
40	145	93	40	0.584	0.200
45	132	83	45	0.627	0.239
50	118	71	50	0.661	0.277
55	105	57	55	0.687	0.311
60	96	39	60	0.704	0.338
65	89	23	65	0.714	0.356
70	90	12	70	0.718	0.367
75	94	6	75	0.720	0.373
80	99	1	80	0.721	0.375
85	101	0	85	0.721	0.375
90	102	0	90	0.721	0.375
95	101	0			
100	96	0			
105	93	0			
110	94	0			
115	96	0			
120	95	0			
125	89	0			
130	82	0			
135	72	0			
140	63	0			
145	54	0			
150	44	0			
155	34	0			
160	22	0			
165	10	0			
170	5	0			
175	0	0			
180	0	0			

Luminance distribution (cd/m² 1000 lm)		
Angle (degrees)	*Transverse plane (T)*	*Axial plane (A)*
45	553.3	606.9
50	502.9	571.1
55	458.7	513.8
60	433.6	403.3
65	419.3	281.4
70	446.7	181.4
75	497.0	119.9
80	564.4	29.8
85	630.0	0.0

illuminance provided by the installation over time, and hence to determine whether the installation is likely to meet the appropriate design service illuminance criterion recommended.

Precise calculation involves the use of data contained in the CIBSE 1984 Lighting Code and lamp manufacturers' technical data books. As a guide, the lumen output of discharge lamps after 2000 h of use is about 0.92 of their initial lumens. Typical depreciation curves for luminaires are shown in Fig. 7.1. This also includes data on light loss due to dirt on the room surfaces.

Utilisation factors UF[*F*]			SHR NOM = 1.50								
Room reflectances			Room index								
C	W	F	0.75	1.00	1.25	1.50	2.00	2.50	3.00	4.00	5.00
0.70	0.50	0.20	0.41	0.47	0.52	0.55	0.60	0.63	0.66	0.69	0.71
	0.30		0.36	0.42	0.47	0.50	0.56	0.59	0.62	0.66	0.68
	0.10		0.32	0.38	0.43	0.47	0.52	0.56	0.59	0.63	0.66
0.50	0.50	0.20	0.37	0.42	0.46	0.49	0.53	0.55	0.57	0.60	0.61
	0.30		0.33	0.38	0.42	0.45	0.49	0.52	0.55	0.57	0.59
	0.10		0.29	0.34	0.39	0.42	0.47	0.50	0.52	0.56	0.58
0.30	0.50	0.20	0.33	0.37	0.40	0.43	0.46	0.48	0.49	0.51	0.53
	0.30		0.29	0.34	0.37	0.40	0.43	0.46	0.48	0.50	0.51
	0.10		0.27	0.31	0.35	0.38	0.41	0.44	0.46	0.48	0.50
0.00	0.00	0.00	0.23	0.26	0.28	0.30	0.33	0.35	0.36	0.38	0.39
BZ class			3	4	4	4	4	4	4	4	5

CIE Flux Code 52/80/92/58/79

Flux fraction ratio 0.72

Conversion terms

Luminaire length (mm)	600	1200	1500	1800	2400
Wattage (W)	1 × 20	1 × 40	1 × 65	1 × 75	1 × 125
Conversion factors (PC and UF)	1.03	1.09	1.03	1.00	1.05

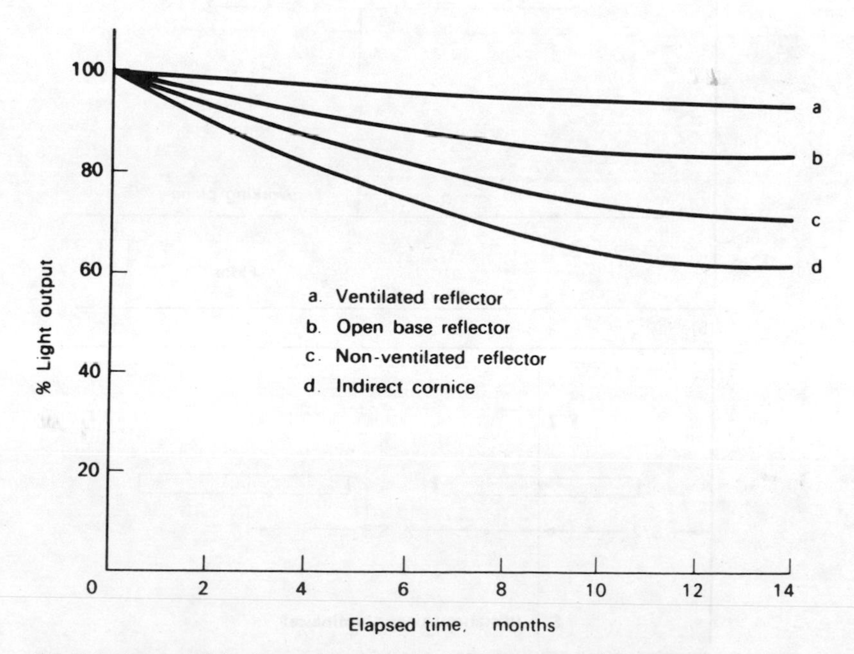

Fig. 7.1 Loss of light due to dirt on luminaires and room surfaces

Example
What is the light loss factor due to open base fluorescent reflector luminaires assuming a 6 month cleaning interval based on lamp lumen output after 2000 h?

Solution
Lamp lumen maintenance factor = 0.92 (from lamp data)
Average 3 month depreciation = 0.90 (from Fig. 7.1)
∴ Light loss factor = 0.92 × 0.90
= 0.83

It must be stressed that this is an over simplification, and that to obtain an accurate factor the Guide must be used.

Planning the layout

The spacing of the luminaires should be such that the ratio of the minimum to the average illuminance over an unobstructed task area should be not less than 0.8. There are also recommendations on the ratios of illuminances on the working plane, walls, and ceilings which will be discussed later in the chapter.

The required uniformity should be achieved if the layout conforms to the manufacturers' quoted maximum spacing to height ratio (S/h_m). The values of S and h_m are illustrated in Fig. 7.2. Where manufacturers' data are not available, the BZ classification

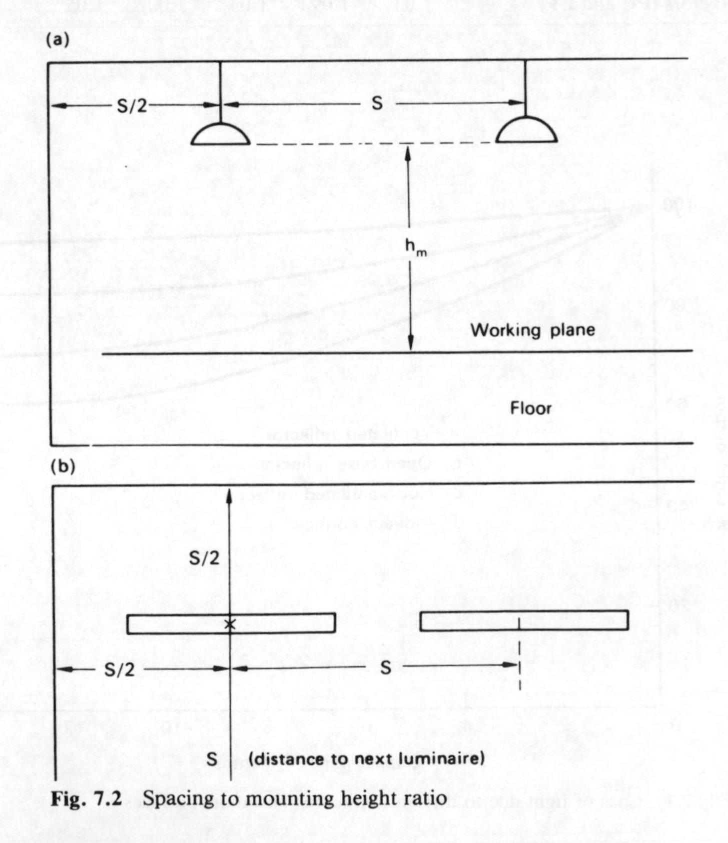

Fig. 7.2 Spacing to mounting height ratio

Table 7.4 Spacing to height ratio for BZ classifications

BZ classification	*Maximum spacing to height ratio*
10–5	1.5 : 1.0
4/3	1.25 : 1.0
2/1	1.0 : 1.0

gives a guide to the ratio. This is shown in Table 7.4. If linear luminaires are used in continuous rows, then the manufacturer may quote a maximum transverse spacing to height ratio. This indicates the maximum spacing between parallel rows of luminaires.

Example

A lighting scheme is required for a small drawing office, 6 m square, with a 3 m ceiling height. Luminaires are to be used whose photometric data are given in Table 7.4.

The office is air conditioned and the room reflectances are 0.7 ceiling, 0.3 walls, and 0.2 floor.

Solution

Step 1. Select the mounting height. In this case the ceiling is relatively low and the luminaires would be mounted direct to the ceiling.

$$\text{Mounting height } h_m = \text{Ceiling height} - \text{Desk height}$$
$$= 3.0 - 0.8 \text{ m}$$
$$= 2.2 \text{ m}$$

Step 2. Determine the illuminance and, hence, the lumens required. The illuminance would normally be selected from the CIBSE schedule. However, a drawing office is an example given in Table 7.1 and the value is 750 lx. Using eqn [7.1], values of UF and LLF must be determined. The room index (eqn [7.2]), is given by:

$$\text{RI} = \frac{6 \times 6}{12 \times 2.2}$$
$$= 1.4$$

from Table 7.3

$$\text{UF} = 0.49$$

and as the office is air conditioned, an LLF value of 0.85 will be assumed.

$$\therefore \text{ Lamp lumens} = \frac{750 \times 6 \times 6}{0.49 \times 0.85}$$
$$= 64\,826 \text{ lm}$$

Step 3. Determine the minimum number of lighting points. This may not produce a satisfactory layout but the golden rule is that this number can be increased *but not reduced.*

The luminaire has a quoted maximum S/h_m ratio of 1.92. If $h_m = 2.2$ m, then

$$S = 2.2 \times 1.92$$
$$= 4.2 \text{ m}$$

So the minimum number of points is 2 × 2, i.e. 4.

Step 4. Can this number achieve 750 lx with a sensible lamping? The available lamp sizes are shown in Table 7.3 and range from a 20 W to a 125 W fluorescent lamp. (A range of typical lumen outputs is given in the appendix on p. 184).

For a layout of four points:

$$\text{Lumens per point} = \frac{64\ 826\ \text{lm}}{4}$$

$$= 16\ 206\ \text{lm}$$

No single lamp approaches this initial lumen value. The nearest is the 125 W white tube with 9500 lm. A possible solution is to use twin lamp luminaires, but a new set of calculations is required. As it is a drawing office it would be more advantageous to increase the number of points to eight or nine to reduce the risk of shadow. A suggested layout is shown in Fig. 7.3.

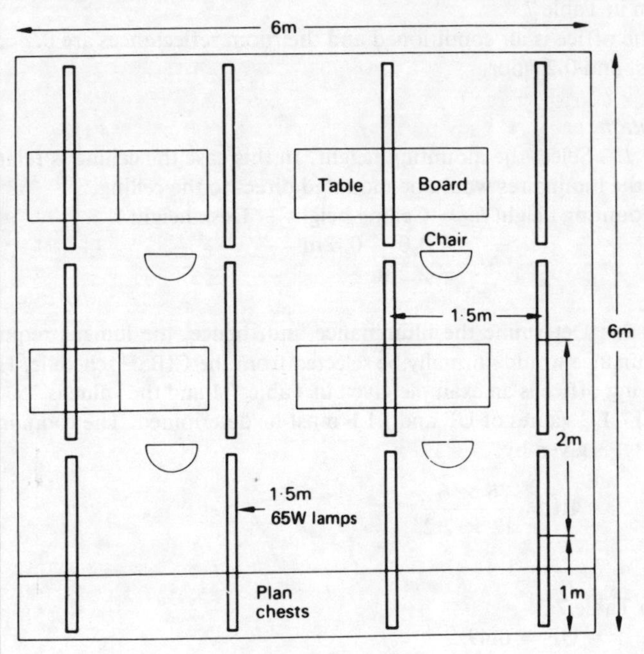

Fig. 7.3 Lighting layout in a drawing office

Types of layouts

General lighting

The simplest and often the most effective layout is the one already described. A regular array of luminaires is planned to achieve an overall level of illuminance. This provides sufficient light for the specified visual task to be carried out anywhere within the room. The pattern of lights can be varied, but in general it is better to keep the lines of luminaires parallel to the room major axis. Diagonal and 'herring bone' patterns can cause a disturbing visual effect.

There are objections to this type of layout, some of which are:

1. The lighting is monotonous and flat. This normally refers to the modelling or degree of shadow, and can be related to the ratio of the vertical and horizontal illuminances. The amount of shadow is, to some extent, a function of the light distribution, and luminaires with a greater degree of downward light (lower BZ classifications) will produce deeper shadows. A general diffusing luminaire (BZ 6–8) will produce a much flatter lighting effect.
2. The lighting has no sense of flow. In daytime, with side windows, there is a sense of direction of light from the windows. This is absent at night, although it could be achieved by using angled luminaires. The CIBSE 1984 Lighting Code suggests:

   (a) the flow of light, in any one part of an interior should be more noticeable in one direction than any other;
   (b) where possible, the dominant direction of flow should be at 30° or more from the vertical, and the direction of flow may be varied in different parts of the space.

   Achievement of these aims is not difficult when combining daylight and electric light, but a very sophisticated lighting scheme is needed to achieve it using electric lighting only.
3. The layout is extravagant. This implies that there is more light than is necessary in certain parts of the room. A minimum design level for normal working interiors is 200 lx, and if the task requires 750 lx, but is performed in only certain parts of the interior, then it is reasonable to provide an overall layout giving 200 lx with localised additional lighting to provide 750 lx. This is best achieved if the installation is flexible, e.g. it is mounted on an electrical trunking system so that luminaires can be moved if the working area is changed.

Local lighting

There are certain visual tasks which can never be adequately lit by general lighting. The obstructions of the machine may obscure the lighting, or visibility may depend on the revealing of texture by shadow which requires the use of an adjustable small source of light, or a very high local illumination may be needed.

When local lights are used, the lamps must be shaded both from the view of the operator and from people in adjacent work areas.

Localised lighting

Where the furniture or workspace layout is fixed and is likely to remain so, the lighting can be related. An example would be a library which uses localised lighting between the book stacks with, perhaps, local lighting on the reading tables, and general lighting in the main circulation area.

The advantages are an improved appearance and more effective lighting. A disadvantage is the lack of flexibility in the use of the room.

Luminous ceilings

The expanding use of suspended ceilings, which provide both an attractive ceiling finish and useful 'cupboard' space in the ceiling void for the environmental ducts, has led to the development of a wide range of luminous ceiling systems. The translucent ceiling can be in the form of plastic diffusing panels or open cells or louvres, and a typical ceiling is shown in Fig. 7.4. The calculations involve the determination of a suitable utilisation factor and the correct spacing of lamps above the ceiling to obtain an even brightness of the

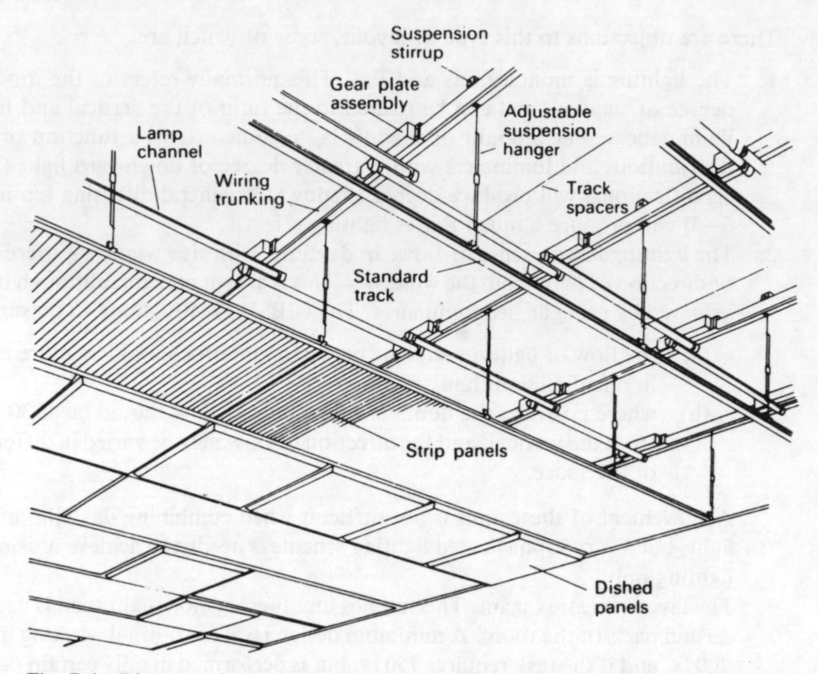

Fig. 7.4 Diagram of a suspended luminous ceiling system (courtesy Courtney, Pope Lighting Ltd)

luminous panels. Discomfort glare cannot be calculated by the normal technique used for individual luminaires.

Illuminance calculations The void of a luminous ceiling system acts like an integrator – i.e. the light is reflected from surface to surface – and provided all surfaces in the void, including pipes and ducts, have high reflectance, the system will be quite efficient whatever the transmittance of the ceiling panels.

Utilisation factors are given in Table 7.5 for a double-skin stretched vinyl film ceiling, with 78 per cent transmittance and lamps in a 0.3 m deep cavity of 0.75 average reflectance.

Louvre type ceilings will, in general, only be more efficient for room indices below 1.0.

In practice, a ceiling void reflectance of 0.75 may be quite impractical, the true value being less than 0.3. This reduces the efficiency considerably and it is necessary to either provide separate reflectors for the lamps or use lamps with internal reflector coats.

Table 7.5 Utilisation factors for a luminous diffusing ceiling

Room index	*0.6*	*1.0*	*1.5*	*2.0*
0.5 Wall reflectance	0.22	0.34	0.41	0.46
0.3 Wall reflectance	0.18	0.30	0.37	0.42

Lamp spacing The intention, normally, is that the position of the lamp should not be apparent. To achieve a uniform appearance with most types of luminous ceiling the spacing (S) to height (h) ratio in the cavity should not be greater than 3/2. This can be extended in a ceiling cavity of high reflectance, or with certain louvred ceilings.

Scalar illuminance

Recommended lighting levels are normally in terms of horizontal illuminance. It is arguable that the restriction to the horizontal plane is unrealistic, as the visual field is in all planes and lighting from the side, such as from windows, may give a comparatively low illuminance on horizontal planes but a much higher value on vertical planes.

The scalar illuminance is the amount of flux falling on a sphere of unit surface area at a point in space. The unit is the lux. In the simplest case, if light were flowing in a downward vertical direction, then E_h at a point would be the incident lumens per unit horizontal area.

If this area were a disc of radius r metres, then

$$E_h = \frac{\text{Flux}}{\pi r^2} \text{lx}$$

The spherical or scalar illuminance at the same point would be

$$E_s = \frac{\text{Flux}}{\text{Area of sphere radius } r}$$

$$= \frac{\text{Flux}}{4 \pi r^2} \text{lx}$$

or

$$E_h = 4E_s \qquad \ldots [7.4]$$

These two values are illustrated in Fig. 7.5.

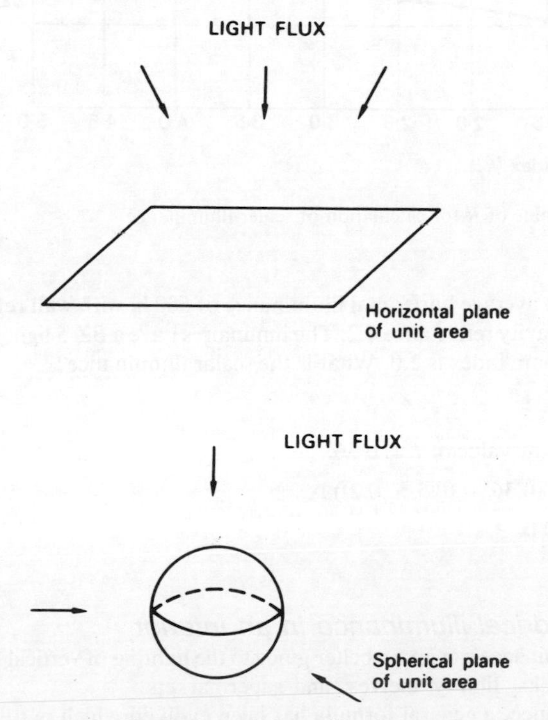

Fig. 7.5 The horizontal planar illuminance (E_h and scalar illuminance (E_s)

Calculation of scalar illuminance in an interior

Scalar values are specified for areas such as corridors. The calculation is a fairly simple extension of that required to calculate the horizontal illuminance (E_h). The value is given by:

$$E_s = E_h(K + 0.5\rho_f)\ \text{lx} \qquad \ldots [7.5]$$

where K is a value depending on the BZ distribution and room reflectances and ρ_f is the average reflectance of the floor cavity. Figure 7.6 gives values of K for medium room reflectance.

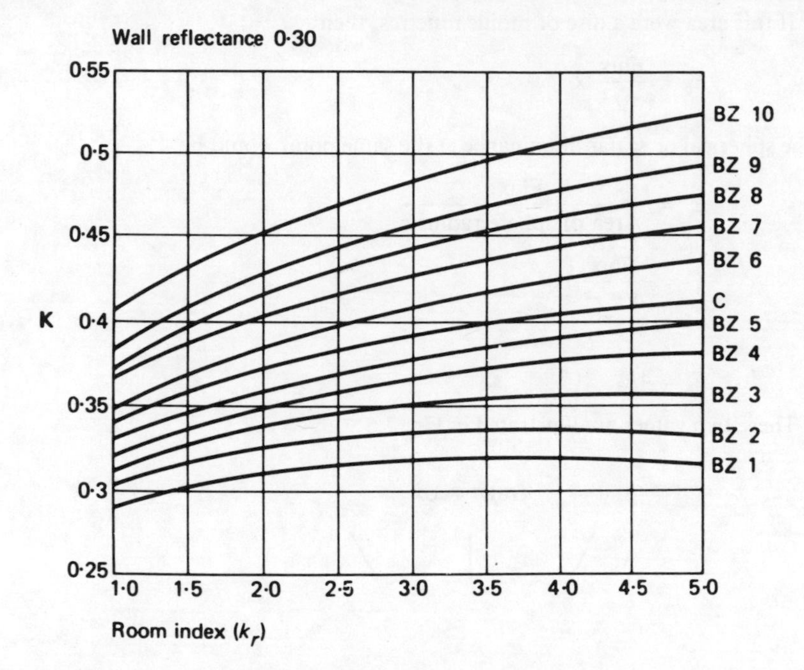

Fig. 7.6 The value of K for calculation of scalar illuminance

Example

A room has an average horizontal illuminance of 500 lx with wall reflectance 0.3 and floor cavity reflectance 0.2. The luminaires have a BZ 5 light distribution and the room index is 2.0. What is the scalar illuminance?

Solution

From Fig. 7.6 the value of K is 0.36.

$$\therefore\ E_s = 500(0.36 + (0.5 \times 0.2))\ \text{lx}$$

$$= 230\ \text{lx}$$

Calculation of cylindrical illuminance in an interior

Cylindrical values of illuminance can give a better guide to the lighting of vertical surfaces. Examples are storage racks, library shelves, and supermarkets.

As with scalar illuminance, a general formula has been evolved which is sufficiently accurate for most purposes.

$$E_c = E_h\,(1.5K - 0.25 + 0.5\,\rho_f)\ \text{lx}$$

K is obtainable from Fig. 7.6 and ρ_f is the reflectance of the floor cavity.

Relative luminance of room surfaces

Calculations of illuminance on the working plane ensure that there will be sufficient light for the task. They do not indicate the general luminance pattern in the room. To find this, the relative illuminances on the working planes, walls, and ceiling must be calculated. Combining these with a knowledge of the surface reflectances enables limited guidance to be made on an acceptable range of luminances. These are illustrated in Fig. 7.7. The figure displays a wide range, and while it is possible to calculate relative illuminances to a fair degree of accuracy, simpler, less precise methods are more likely to be used.

Table 7.6 from the CIBSE 1984 Lighting Code is provided to act as an aid in the selection of a lighting system. It includes suggested patterns of room brightness in a diagram whose explanation is given in Table 7.7. This indicates the anticipated relative brightness of the ceiling and walls in a room with average surface reflectances for that specific type of luminaire.

Approximate method of calculating room illuminance ratios

The following data are required:

(1) the reflectance and size of room surfaces;
(2) the ULOR and DLOR of the luminaires;
(3) the direct ratio of the luminaires, which is

$$\left(\frac{\text{Downward light of installation falling direct on working plane}}{\text{Total downward light}}\right) \quad \ldots [7.6]$$

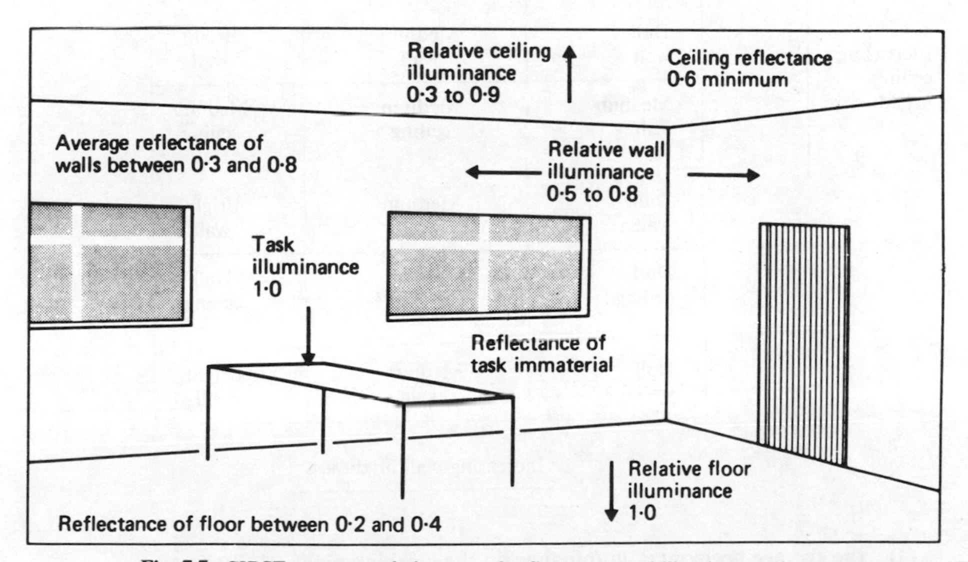

Fig. 7.7 CIBSE recommended range of reflectances and illuminances for room surfaces (reproduced by permission of the Chartered Institution of Building Services Engineers. CIBSE 1984 Lighting Code)

Table 7.6 Typical luminaire characteristics (excerpt) (reproduced by permission of the Chartered Institution of Building Services Engineers. CIBSE 1984 Lighting Code)

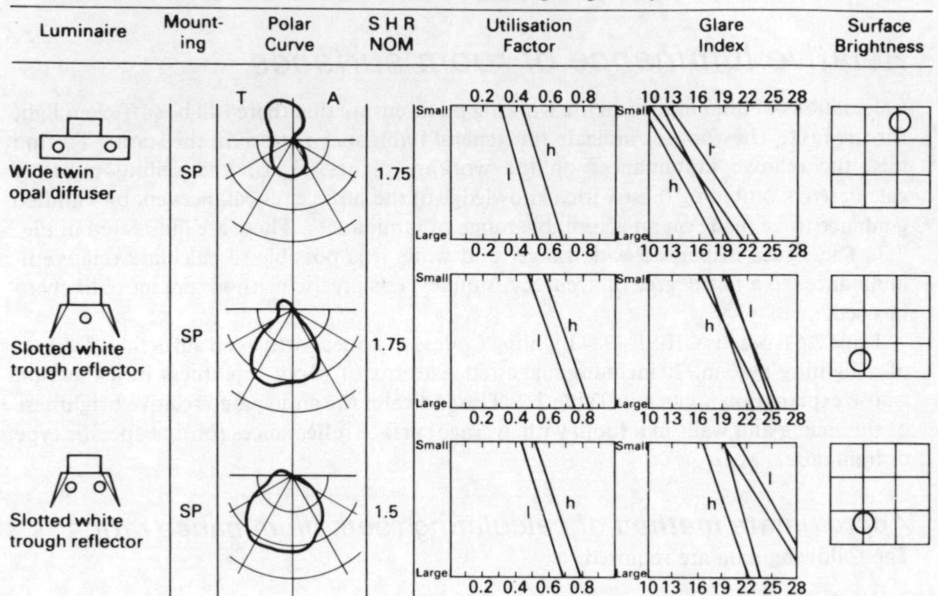

Luminaire	Mounting	Polar Curve	S H R NOM	Utilisation Factor	Glare Index	Surface Brightness
Wide twin opal diffuser	SP		1.75			
Slotted white trough reflector	SP		1.75			
Slotted white trough reflector	SP		1.5			

Table 7.7 Key to pattern of room surface in brightness in Table 7.6 (from CIBSE Code for Interior Lighting, 1984)

Increasing ceiling brightness ↑			
	Bright ceiling Dull wall	Bright ceiling Medium wall	Bright ceiling Bright wall
	Medium ceiling Dull wall	Medium ceiling Medium wall	Medium ceiling Bright wall
	Dull ceiling Dull wall	Dull ceiling Medium wall	Dull ceiling Bright wall

Increasing wall brightness →

(4) the average horizontal illuminance on the working plane (floor) (E_h);
(5) the initial light output of all lamps (F), lm.

The luminance of the room surfaces will comprise:

Direct illuminance (from luminaires) + Reflected illuminance (from other surfaces) × Reflectance of surface.

The direct illuminance on the working (E dwp). This is given by the formula

$$E_{dwp} = \frac{E \times DLOR \times DR \times LLF}{\text{Area of working plane}} \qquad \ldots [7.7]$$

The direct illuminance on walls (E dn). This will be a function of the downward flux not reaching the working plane directly:

$$E_{dw} = \frac{E \times DLOR \times (1 - DR) \times LLF}{\text{Area of four walls}} \qquad \ldots [7.8]$$

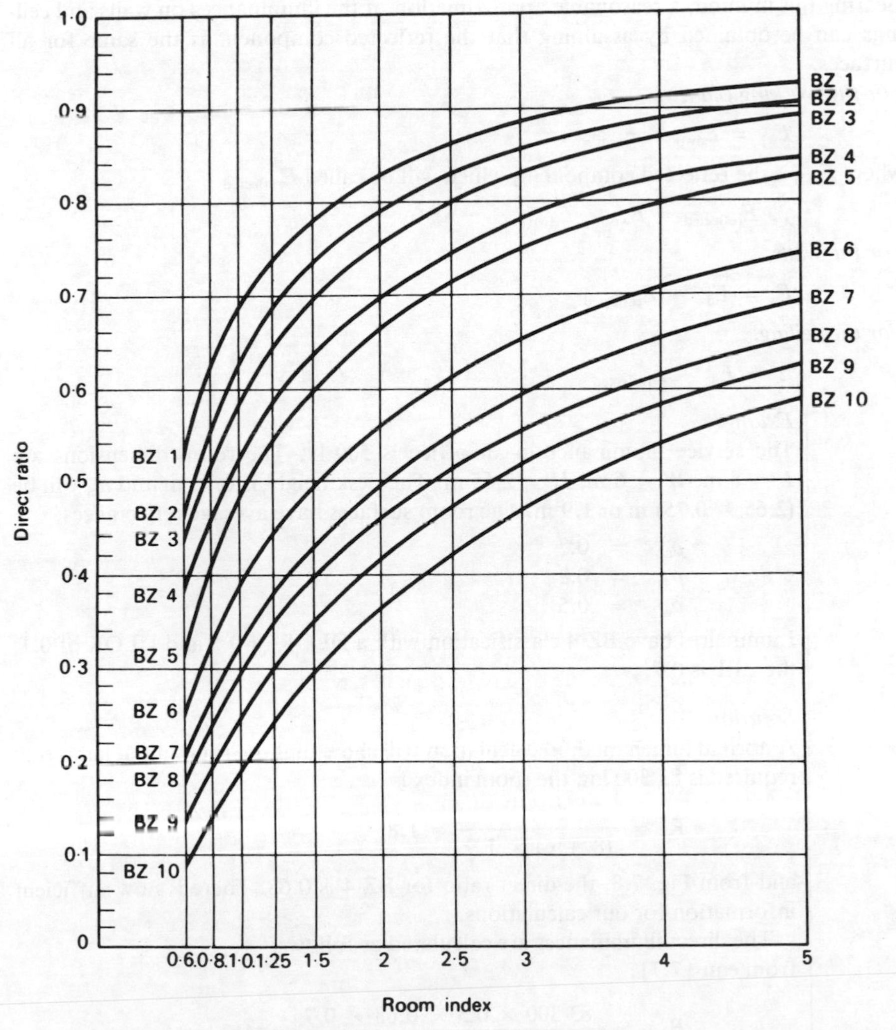

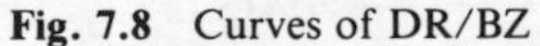

Fig. 7.8 Curves of DR/BZ

The direct illuminance on the ceilings (E dc). The upward flux will all fall directly onto the ceiling cavity:

$$E_{dc} = \frac{E \times \text{ULOR} \times \text{LLF}}{\text{Area of ceiling cavity}} \qquad \ldots [7.9]$$

Each room surface now acts as a light source and provides the indirect or reflected illuminance component. The total illuminance on each surface will be the sum of the direct and reflected components.

E_{wp} = Horizontal illuminance originally calculated for the lighting scheme

The next step is to calculate the reflected component of illuminance on the walls and ceiling. Before making precise calculations it is relevant to look at the recommendations. These are usually quoted in terms of a range of reflectances and illuminances, since luminance is proportional to illuminance times reflectance.

It is unrealistic to quote precise values as a wide range of luminances is acceptable. Bearing this in mind, a reasonable approximation of the illuminances on walls and ceilings can be obtained by assuming that the reflected component is the same for all surfaces.

For the working plane:

$$E_{wp} = E_{dwp} + E_{rwp}$$

where E_{rwp} is the reflected component, which will be called $E_{reflected}$

$$\therefore\ E_{reflected} = E_{wp} - E_{dwp}$$

For the walls:

$$E_{w} = E_{dw} + E_{reflected}$$

For the ceiling:

$$E_{c} = E_{dc} + E_{reflected}$$

Example

The service illuminance in an office is 500 lx. The room dimensions are L = 8 m, W = 6 m, H = 2.65 m. The desk height is 0.75 m and h_m will be (2.65 − 0.75) m or 1.9 m. The room surfaces have average reflectances:

$$\rho_c = 0.7$$
$$\rho_{wp} = 0.3$$
$$\rho_w = 0.5$$

Luminaires have BZ 4 classification with a DLOR of 0.5 and ULOR of 0.1; the LLF is 0.7.

Solution

A normal lumen method calculation will show that the total initial lamp flux required is 83 300 lm; the room index is

$$\text{RI} = \frac{8 \times 6}{(8 + 6) \times 1.9} = 1.8$$

and from Fig. 7.8, the direct ratio for BZ 4 is 0.68. There is now sufficient information for our calculations.

The direct illuminances are calculated as follows:
from eqn [7.7],

$$E_{dwp} = \frac{83\ 300 \times 0.5 \times 0.68 \times 0.7}{8 \times 6} = 413 \text{ lx}$$

from eqn [7.8],

$$E_{dw} = \frac{83\,300 \times 0.5 \times 0.32 \times 0.7}{1.9 \times 2 \times (8 + 6)} = 175 \text{ lx}$$

from eqn [7.9],

$$E_{dc} = \frac{83\,300 \times 0.1 \times 0.7}{8 \times 6} = 121 \text{ lx}$$

For the working plane:

$E_{reflected}$ = 500 − 413 lx
= 87 lx

$\therefore\ E_w$ = 176 + 87 lx
= 263 lx

E_c = 121 + 87 lx
= 208 lx

Referring to Fig. 7.7:

$$\text{Wall illuminance} = \frac{262}{500}\text{ Floor illuminance}$$

= 0.52 (which is just acceptable)

$$\text{Ceiling illuminance} = \frac{208}{500}\text{ Floor illuminance}$$

= 0.41 (which is acceptable)

The luminance values are given by

$$\text{Luminance} = \frac{\text{Illuminance} \times \text{Reflectance}}{\pi}\ \text{cd/m}^2$$

∴ For the working plane:

$$\text{Luminance} = \frac{500 \times 0.3}{\pi} = 48 \text{ cd/m}^2$$

For the walls:

$$\text{Luminance} = \frac{263 \times 0.5}{\pi} = 41.8 \text{ cd/m}^2$$

For the ceiling:

$$\text{Luminance} = \frac{208 \times 0.7}{\pi} = 46.3 \text{ cd/m}^2$$

It must be emphasised that this is an approximate method devised by the author. More precise methods can be found in the CIBSE 1984 Lighting Code, Technical Memorandum No. 5 and Technical Report No. 15.

Calculation of glare

In both daylight and electric light there is always a need to restrict the luminances within the normal field of view, otherwise an unacceptable degree of glare will be experienced.

Disability glare

This can be defined as: *Glare which impairs the ability to see detail without necessarily causing visual discomfort.* As it is a definite reduction in the ability to see, it can be measured in terms of visual performance, and its effect can be expressed as a shift in the adaptation level of the eye (see Chapter 1). If the adaptation level is raised by a light source, such as an unshaded lamp or window, then the eye becomes less sensitive to small differences in brightness.

One formula for expressing this change in adaptation level is:

$$\text{Adaptation luminance} = L_0 + \frac{kE}{\theta^2} \qquad \ldots [7.10]$$

where L_0 is the original average luminance of the field of view
E is the illuminance from glare source at the eye
θ is the angle between the line of sight and the direction of the glare source
k is a constant related to the age of the observer.

Present attempts to identify and design for the avoidance of disability glare in the task are based on the work on contrast rendition factors (see Chapter 1). Details are contained in the CIE Publication No. 19: *A unified framework of methods for evaluating visual performance aspects of lighting*. There are subsequent papers in the UK, principally by the Building Research Establishment and the Electricity Council, but the stage has not yet been reached when specific recommendations can be made and accepted.

Discomfort glare

This can be defined as: *Glare which causes visual discomfort without necessarily impairing the vision.* The degree of discomfort will depend on the type of location and angle of view; for instance, people will tolerate a much brighter installation in a supermarket, where they are continually on the move and looking in all directions, than in a classroom, where the direction of view is fixed, the visual task more exacting, and the occupants have more time to become aware of the lighting. This is very much a subjective assessment; it cannot be measured in terms of performance, but only in relative degrees of discomfort. Work at the Building Research Establishment has provided the basic data for the CIBSE system of classifying discomfort glare, known as the *Glare Index System*. Glare from a light source can be expressed as:

$$\text{Discomfort glare } (g) = \frac{L_s^{1.6} \times \omega^{0.8}}{L_B \times P^{1.6}} \times 0.478 \qquad \ldots [7.11]$$

where L_s is the luminance of the source,
L_B is the average luminance of the background,
ω is the angular size of the source,
P is the position index which indicates the effect of the position of the source on glare.

To calculate the glare from each light point for a range of room positions would be too time-consuming. The method can be simplified by considering a number of standard conditions and making adjustments as necessary. The standard conditions and assumptions are:

1. Glare is additive; therefore, for a complete installation, the glare from all the luminaires would be the same as for a single luminaire of the same luminance L_s but whose angular size ω is the total of all the individual luminaires.
2. The direction of view is horizontal, straight across the room, and 1.2 m above floor level.
3. The room is rectangular and has a regular array of luminaires.

The calculations and corrections depend on the tabular data being used. The broad steps are:

1. Find the initial glare index for the particular type of luminaire in a specific room under standard conditions. The index is 10 log g, where g, the discomfort glare, can be found from [7.11].
2. Apply various correction factors as required by the system being used. this will give the final glare index. This value should not exceed the limiting glare index recommended for that particular type of location. A list of typical limiting glare indices is given in Table 7.8.

Table 7.8 CIBSE recommended limiting glare indices for different classes of visual task

Class of visual task	*Examples*	*Limiting glare index*
Critical, often with a fixed direction of view	Drawing offices; very fine inspection	16
Critical, but general direction of view	Offices; libraries; computer buildings	19
Ordinary task with general direction of view	Kitchens; reception areas; fine assembly work	22
Large task or limited viewing time	Stock rooms; medium assembly work	25
No specific visual task or direction of view	Rough industrial work; indoor parking areas	28

The full method is given in CIBSE Technical Memorandum No. 10. It is fairly laborious and some manufacturers now produce glare data for specific luminaires.

Table 7.9 shows a typical glare data sheet which, in this case, relates to the luminaire data given in Table 7.3. The glare data sheet is based on a number of assumptions:

(1) the luminaire is 2 m above eye level;
(2) the lamp emits 1000 lumens;
(3) the spacing to height ratio is 1 : 1.

Table 7.10 gives corrections for factors (1) and (2). No corrections are available for (3).

Initial glare indices are tabulated according to room dimensions and reflectances. Figure 7.9 shows the method of specifying the room dimensions. The Y dimension is always parallel to the line of sight and the X dimension is perpendicular to the line of sight. They are both expressed as multiples of the mounting height above eye level.

One view of the room will show the ends of the luminaires (endwise view) and the other view will show the sides (crosswise view).

Table 7.9 Glare data for Thorn/EMI prismatic enclosure type luminaire (see Table 7.3 for general photometric data)

Glare indices											
Ceiling reflectance		0.70	0.70	0.50	0.50	0.30	0.70	0.70	0.50	0.50	0.30
Wall reflectance		0.50	0.30	0.50	0.30	0.30	0.50	0.30	0.50	0.30	0.30
Floor reflectance		0.14	0.14	0.14	0.14	0.14	0.14	0.14	0.14	0.14	0.14
Room dimension		*Viewed crosswise*					*Viewed endwise*				
X	*Y*										
2*H*	2*H*	7.3	8.5	8.4	9.7	11.1	6.7	8.0	7.8	9.1	10.6
	3*H*	9.4	10.5	10.5	11.6	13.1	8.4	9.5	9.5	10.7	12.2
	4*H*	10.4	11.4	11.5	12.6	14.1	9.1	10.2	10.3	11.4	12.9
	6*H*	11.5	12.5	12.6	13.6	15.2	9.8	10.7	10.9	11.9	13.5
	8*H*	12.1	13.0	13.3	14.2	15.8	10.0	10.9	11.1	12.1	13.7
	12*H*	12.8	13.7	13.9	14.9	16.4	10.1	11.0	11.2	12.2	13.7
4*H*	2*H*	8.1	9.1	9.2	10.3	11.8	7.6	8.7	8.8	9.8	11.4
	3*H*	10.5	11.4	11.6	12.6	14.1	9.6	10.5	10.8	11.7	13.2
	4*H*	11.7	12.5	12.8	13.7	15.3	10.5	11.3	11.7	12.5	14.1
	6*H*	13.0	13.8	14.2	15.0	16.6	11.4	12.1	12.6	13.3	14.9
	8*H*	13.8	14.5	15.0	15.7	17.3	11.7	12.4	12.9	13.6	15.2
	12*H*	14.6	15.2	15.8	16.4	18.1	11.9	12.5	13.1	13.8	15.4
8*H*	4*H*	12.3	13.0	13.5	14.2	15.8	11.3	12.0	12.5	13.2	14.8
	6*H*	13.9	14.5	15.2	15.7	17.4	12.5	13.1	13.7	14.3	15.9
	8*H*	14.9	15.4	16.1	16.6	18.2	13.0	13.5	14.2	14.7	16.4
	12*H*	15.9	16.4	17.2	17.6	19.3	13.4	13.8	14.6	15.1	16.7
12*H*	4*H*	12.4	13.0	13.6	14.2	15.8	11.6	12.2	12.8	13.4	15.0
	6*H*	14.1	14.7	15.4	15.9	17.5	12.9	13.4	14.1	14.6	16.2
	8*H*	15.2	15.7	16.5	16.9	18.6	13.5	14.0	14.7	15.2	16.9
	12*H*	16.1	16.5	17.4	17.8	19.4	13.7	14.1	15.0	15.4	17.0

Glare index conversion:

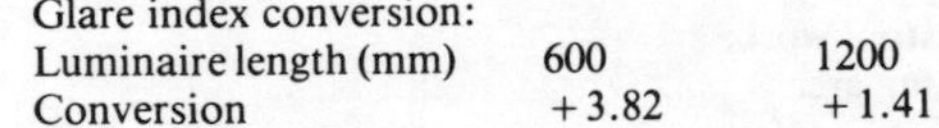
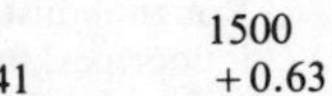
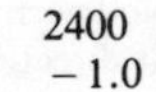

Luminaire length (mm)	600	1200	1500	1800	2400
Conversion	+3.82	+1.41	+0.63	0	−1.0

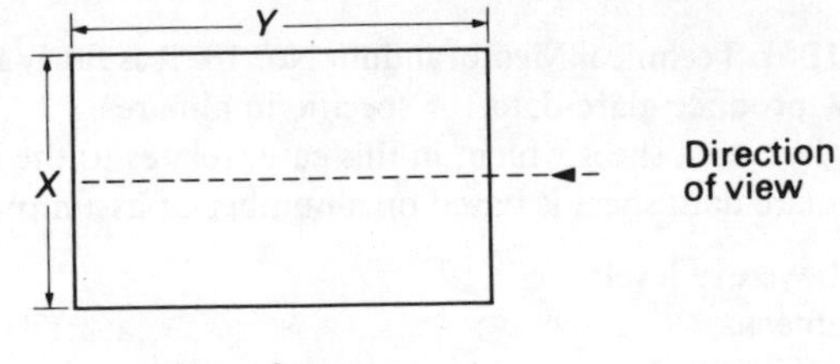

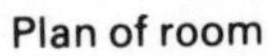

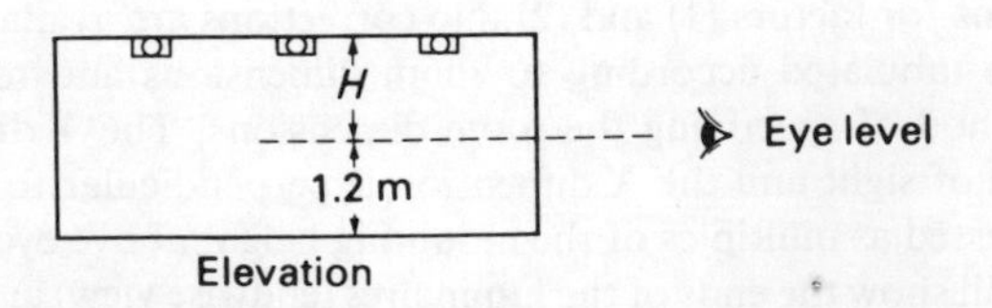

Fig. 7.9 Room dimensions for CIBSE glare calculations

When the initial glare index has been found it must be corrected for:

(1) mounting height above 1.2 m eye level if this differs from 3 m;
(2) total downward luminous flux if this differs from 1000 lm;
(3) extra correction terms if the published glare index table covers a variety of luminaire sizes or lamp types.

These correction terms are added to (or subtracted from) the initial glare index to give the final glare index of the installation.

Example
Referring to the scheme designed earlier in this chapter and illustrated in Fig. 7.3, what is the glare index?

Solution
The room is square, so $X = Y = 6$ m. The ceiling height is 3 m and the height of eye level for a seated man is 1.2 m above the floor.

$$H = 3 - 1.2 \text{ m}$$
$$= 1.8 \text{ m}$$
$$\therefore \quad X = 6 \text{ m}$$
$$= \frac{6H}{1.8}$$
$$= 3.3H = Y$$

The reflectances are 0.7 ceiling, 0.3 wall, and 0.2 floor. In this case the worst glare situation is when the luminaires are viewed crosswise. From Table 7.9, the initial glare index is found by interpolation:

X	Y		$Y = 3.3H$	$X = 3.3H$
$2H$	$3H$	10.5	10.8	
	$4H$	11.4		11.4
$4H$	$3H$	11.4	11.8	
	$4H$	12.5		

Applying corrections from Table 7.10:

Initial lamp lumens = 5100 lm
Correction = +4.2

Height above eye level is 1.8 m
Correction = −0.2

There is a further glare index conversion of +0.63 (see Table 7.9) for 1.5m lamp
∴ Final glare index = 11.4 + 4.2 −0.2 + 0.63
= 16.03

As the limiting glare index for a drawing office is 16, this installation barely meets the glare limit requirements. It does suggest that glare could be a problem, and another type of luminaire should be selected.

Table 7.10 Correction data for use with Table 7.9

Initial lamp lumens	*Conversion term*	*Height H above 1.2 m eye level (m)*	*Conversion term*
100	−6.0	1	−1.2
150	−4.9	1.5	−0.5
200	−4.2		
300	−3.1	2	0.0
500	−1.8		
700	−0.9	2.5	+0.4
1 000	0.0	3	+0.7
1 500	+1.1	3.5	+1.0
2 000	+1.8		
3 000	+2.9	4	+1.2
5 000	+4.2	5	+1.6
7 000	+5.1	6	+1.9
10 000	+6.0		
15 000	+7.1	8	+2.4
20 000	+7.8	10	+2.8
30 000	+8.9	12	+3.1
50 000	+10.2		

Discomfort glare from large areas

The glare index system refers to layouts of individual luminaires. In the case of large luminous areas, such as luminous diffusing ceilings, it is more effective to control discomfort glare by limiting the source brightness at normal viewing angles. It is recommended that overall diffusing luminous ceilings are not used where a glare index less than 19 is specified. Elsewhere, the luminance should not exceed 500 cd/m².

Proposed CIE luminance curve system

The UK system has its critics and is not internationally accepted. Some consider it too precise and complicated for assessing a subjective limit and the CIE favour a method developed in Germany and Holland. Some UK manufacturers provide data for the CIE method to be used, and, as will be shown, it is relatively simple.

Table 7.11 CIE glare system: relation quality class (A–E) to curve letter (*a*–*h*)

Quality class	*Valid for service illuminance E (lx)*							
A	2000	1000	500	⩽300				
B		2000	1000	500	⩽300			
C			2000	1000	500	⩽300		
D				2000	1000	500	⩽300	
E					2000	1000	500	⩽300
	a	*b*	*c*	*d*	*e*	*f*	*g*	*h*

The degree of glare is assessed in terms of luminances in the axial and transverse planes between 45° and 85°. Table 7.11 and Fig. 7.10 are used to check if the luminaire chosen will be acceptable in a room of specific size for a specific class of interior. Table 7.12 lists the five classes, with examples.

Luminaries described in this system use the following terms:

Luminous sides – luminous side panels with a height greater than 30 mm.

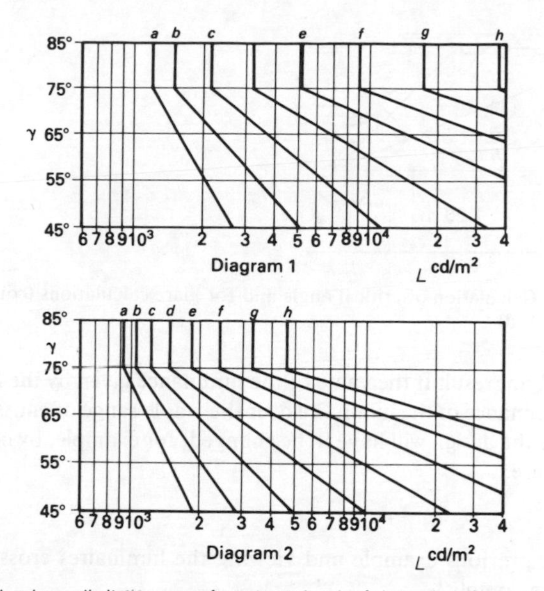

Fig. 7.10 Luminance limitation curves for a stepped scale of glare ratings (0 = no glare, to 6 = intolerable glare) representing quality classes A to E, and for various values of standard service illuminance *E*. (courtesy Philips Electronic Ltd)

Diagram 1: Valid for
- all luminaires without luminous sides
- all elongated luminaires viewed lengthwise

Diagram 2: Valid for
- all non-elongated luminaires with luminous sides
- elongated luminaires with luminous sides when viewed crosswise

Table 7.12 Quality classes of interiors for CIE glare system

Class	*Quality*	*Example*
A	Very high	Drawing office
B	High	General office
C	Medium	Supermarket
D	Low	Toilets
E	Very low	Iron foundry

Elongated – when ratio of length to width of the plan luminous area is not less than 2 : 1.

The steps in the calculation are:

1. Determine the mean luminances between 45° and 85° of the type of luminaire selected for the installation.
2. Determine the quality, class and illuminance level required for the installation in the new condition.
3. Select the appropriate curve (class and level) of the relevant diagram (Fig. 7.10).
4. Determine the maximum angle to be considered for the room length and height between eye level and the plane of the luminaires. To do this take the horizontal line on the glare limitation diagram for the value of a/h_s thus found. The part of the curve above this line may be ignored. The value of a/h_s is found from Fig. 7.11.
5. Compare the luminance of one luminaire with the selected part of the limiting curve.

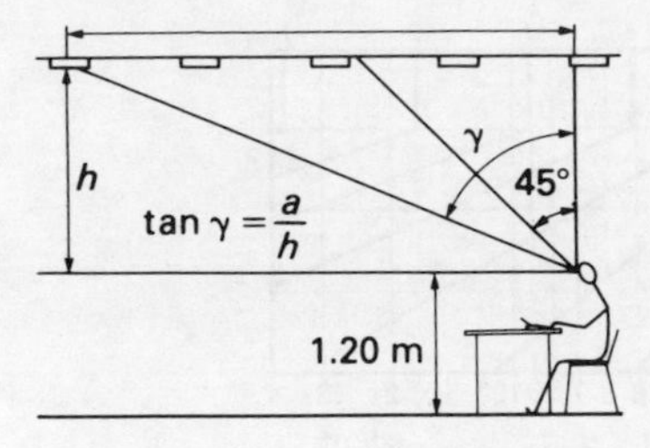

Fig. 7.11 Calculation of critical angle and for glare calculations (courtesy Philips Electronic Ltd)

Discomfort glare will not result if the value of the luminance given by the limiting curve exceeds the actual luminance of the luminaire over the whole range of emission angles. If the result is otherwise the design will have to be changed; for example, by choosing a different type of luminaire.

Example
Using the previous example and viewing the luminaires crosswise, will the installation comply?

Solution
Luminance data is included in Table 7.3. These values are based on 1000 lm and will need correcting to 5100 lm. As this is a drawing office, Class A would be appropriate. The illuminance level is already designed to 750 lx. the luminaire would be classed as elongated, luminous sided.

From Table 7.11 either curve *b* or *c* could be selected for use in Fig. 7.10. The relevant range of angles for luminance values is from 45° upwards. The upper limit (γ) depends on the room shape, and this is illustrated in Fig. 7.11.

In this example h_s is 1.8 m and a is 6 m; hence

$$\tan \gamma = \frac{a}{h_s}$$

$$= \frac{6}{1.8}$$

$$= 73°$$

So only 45° to 73° need be considered. From Table 7.3 for a transverse plane:

$$\text{Luminance at } 45° = 553 \times \frac{5100}{1000} \text{ cd/m}^2$$

$$= 2822 \text{ cd/m}^2$$

$$\text{Luminance at } 55° = 459 \times \frac{5100}{1000} \text{ cd/m}^2$$

$$= 2615 \text{ cd/m}^2$$

$$\text{Luminance at } 65° = 419 \times \frac{5100}{1000} \text{ cd/m}^2$$

$$= 2390 \text{ cd/m}^2$$

$$\text{Luminance at } 75° = 94 \times \frac{5700}{1000} \text{ cd/m}^2$$

$$= 535.8 \text{ cd/m}^2$$

Looking at Fig 7.10, Diagram 2, the 45° value lies on b, the 55° value between b & c, and the 65° value between c & d. As with the CIBSE method this is indicating excessive glare for this luminaire in this type of situation.

Alternative methods are used in USA (visual comfort rating) and in Australia (luminance limit system). Details are given in their respective lighting codes.

Uplighting

Indirect lighting has been hailed as the 'lighting of the 80s'. It was also the 'lighting of the 30s' and one may well think 'so what's new?' As in the 30s, it claims to be glare free. It enables the use of higher lamp wattages than normal with relatively low ceilings, and its revival is closely linked with the computer age. It is still a relatively inefficient method of lighting, highly dependent on a good standard of maintenance and you either like it or you don't.

There are various techniques for planning an uplighter scheme and these notes will deal with two methods and compare their answers.

1. Lumen method using UF tables
2. Thorn method – 'Equivalent Light Source'

1. Lumen Method

Approximate utilisation factors are available for indirect lighting. These can be used, provided the ULOR is known, to give an average illuminance value. What this method cannot do is give an indication of uniformity. It just assumes the lighting will be uniform.

To compare the two methods we will apply each one to a set situation.

Example
An office 8 m × 8 m is to be lit by four 250 W MBIF lamps in uplighters. The following information is available:

Ceiling reflectance	0.7
Wall reflectance	0.5
Mounting height	1.8 m
Ceiling height	2.8 m

Step 1 The ceiling is effectively a cavity, (2.8 – 1.8)m deep. Because the ceiling cavity, i.e. that upper part of the room which received direct light from the uplighter, is not flat, light will be absorbed by inter-reflection within the cavity. The effective reflectance of the ceiling cavity will be less than the actual reflectances of the surfaces. To calculate the effective reflectance (R_E) of the ceiling cavity:

$$R_E = \frac{\text{CI} \times R_A}{\text{CI} + 2(1 - R_A)} \quad \ldots [7.12]$$

CI is the cavity index (like Room Index)
R_A is the average reflectance of the cavity surfaces

$$\therefore \quad CI = \frac{\text{area of flat ceiling}}{\text{area of } \frac{1}{2} \text{ walls of cavity}} \qquad \ldots [7.13]$$

$$= \frac{8 \times 8}{(8 + 8) \times (2.8 - 1.8)}$$

$$= 4$$

$$R_A = \frac{\text{ceiling area} \times \text{reflectance} + \text{cavity wall area} \times \text{reflectance}}{\text{ceiling + cavity wall areas}} \qquad \ldots [7.14]$$

$$= \frac{(8 \times 8) \times 0.7 + [2 \times (8 + 8) \times 1.0] \times 0.5}{(8 \times 8) + 2(8 + 8) \times 1.0}$$

$$= 0.63$$

$$R_E = \frac{3 \times 0.63}{3 + 2(1 - 0.63)}$$

$$= 0.50$$

To calculate room index, RI, take the height from working plane (0.8 m) to luminaire (1.8 m):

$$RI = \frac{8 \times 8}{(8 + 8) \times (1.8 - 0.8)}$$

$$= 4$$

For effective ceiling reflectance 0.5, walls 0.5

UF = 0.26

This is for a ULOR of 0.75 and is found from Table 7.13. Assume a light loss factor of 0.7 (a dangerous thing to do). Initial lumen output of 250 W MBIF – 19 000 lm. Using lumen formula:

$$\text{Illuminance} = \frac{4 \times 19\,000 \times 0.26 \times 0.7}{8 \times 8}$$

$$= 216 \text{ lx}$$

2. Thorn Lighting Method

L Bedocs, J Lynes, and J Hugill presented a paper to the 1984 National Lighting Conference entitled 'Estimating Uplighting'. The main idea was that if the illuminance from a combination of a single point and large ceiling area was measured on a horizontal working plane, it was possible to work backwards to suggest that there was an equivalent direct luminaire that would give the same distribution as the uplighter and *first* reflection off the ceiling. This is illustrated in Fig. 7.12.

The merit of this idea is that the illuminance values for different values of H_0 can

Table 7.13 Utilisation factors for upward lighting

	Typical/Outline	Basic LOR 75%	*Reflectance* Ceiling	*0.7*			*0.5*			*0.3*		
			Walls	*0.5*	*0.3*	*0.1*	*0.5*	*0.3*	*0.1*	*0.5*	*0.3*	*0.1*
			Room Index									
Totally indirect luminaire. Based on Upward Light Output Ratio 75% (Upper and lower walls the same colour)			0.6	0.1	0.07	0.04	0.07	0.05	0.03			
			0.8	0.13	0.11	0.08	0.11	0.09	0.07			
			1	0.16	0.15	0.12	0.15	0.12	0.10			
Suitable for approximate calculations of 'uplighting' schemes			1.25	0.2	0.19	0.16	0.18	0.15	0.13			
			1.5	0.24	0.23	0.2	0.2	0.18	0.16			
			2	0.28	0.27	0.23	0.22	0.2	0.18			
			2.5	0.32	0.31	0.26	0.24	0.22	0.2			
			3	0.36	0.35	0.29	0.25	0.23	0.21			
			4	0.4	0.38	0.31	0.26	0.24	0.22			
			5	0.43	0.4	0.33	0.27	0.25	0.23			

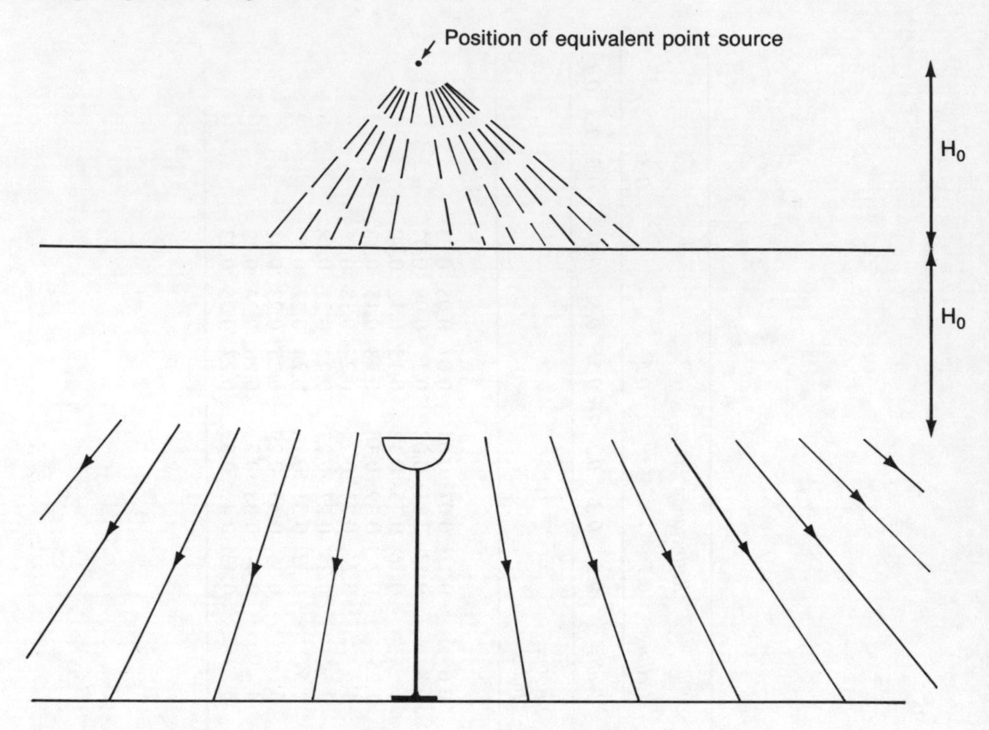

Fig. 7.12 The equivalent point source (courtesy Thorn Lighting)

then be derived. The illuminance due to second and subsequent reflectances are added and this is done with the help of a table.

Photometric data are published for a 2.8 m ceiling height and an example is shown in Fig. 7.13. The graph indicates the horizontal illuminance per 1000 lm at various distances along the working plane.

Example

The design stages are:

1. Work out approximate number of uplighters needed;
2. Calculate the direct illuminances using the illuminance curve;
3. Calculate the inter-reflected component of illuminance;
4. Finally, check the ceiling luminance to see it does not exceed 1500 cd/m^2, with an average value not above 500 cd/m^2.

This may all seem long-winded but it has the merit that it is based on manufacturer's data.

Step 1 – approximate calculation This could be done by the first method but a simple rule of thumb is now provided:

To provide uniform illuminance of 500 lx, a 250 W MBIF lamp covers 15 m^2 and a 250 W SON-DL lamp covers 20 m^2.

In this example there are 4 × 250 W MBIF lamps/64 m^2, so there is 1 point per 16 m^2.

Description : Freestanding circular cantilever uplight

Conversion Terms

TLL Ref	Catalogue Number	Lamp	LOR
UL018	DUGY 250	250W MBIF/SONDL-E	0.84
UL019	DUSY 150	150W SONDL-E	0.85

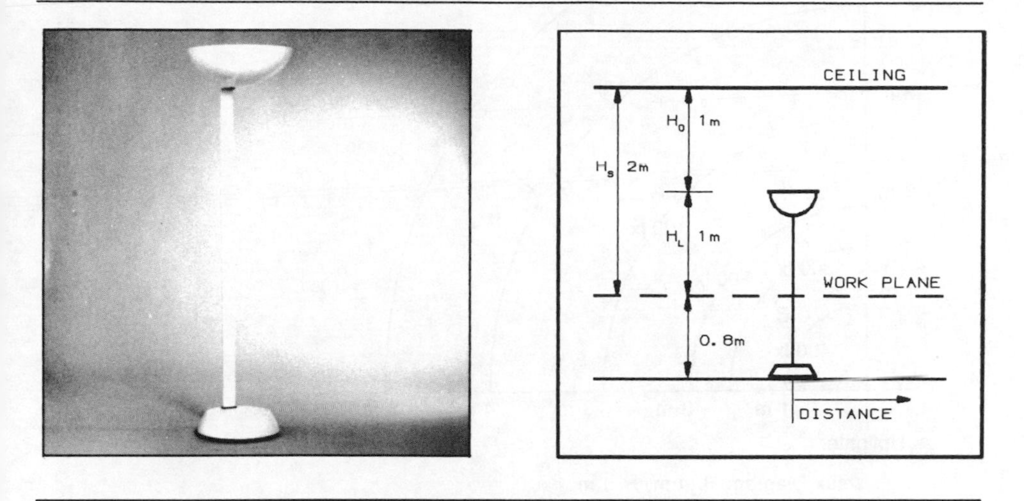

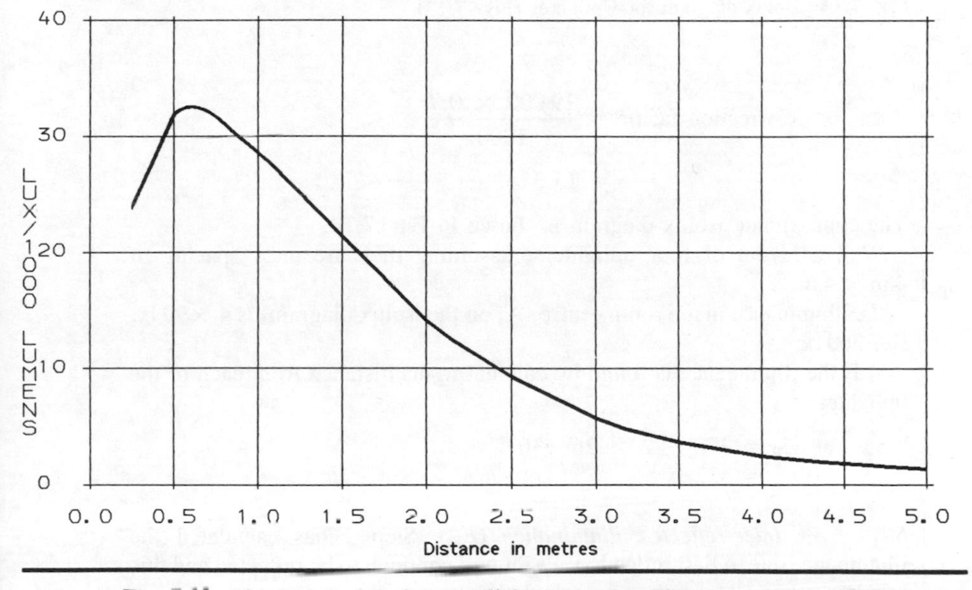

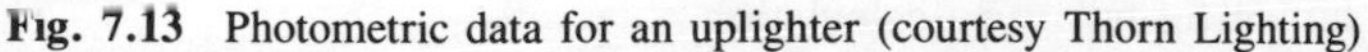

Fig. 7.13 Photometric data for an uplighter (courtesy Thorn Lighting)

Step 2 – direct illuminances An isolux diagram can be drawn for one uplighter. The values must be uprated for lamp lumen correction and a LLF applied.

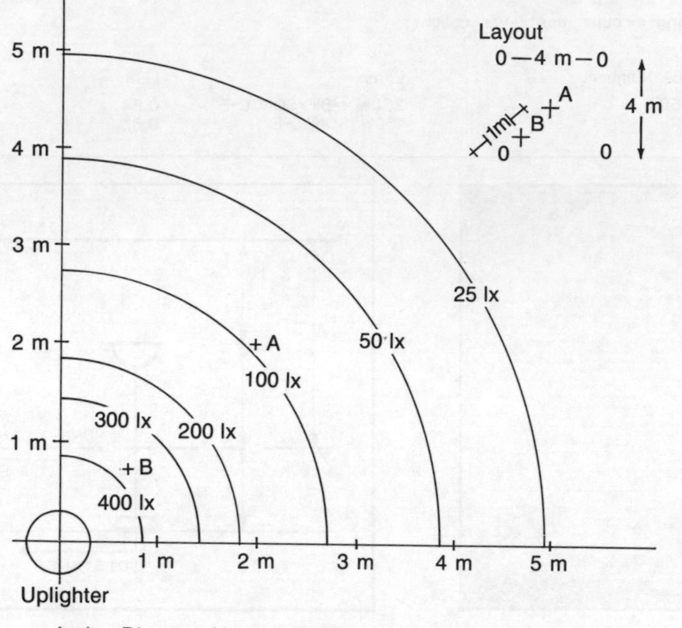

Fig. 7.14 Isolux diagram for Uplighter (Fig. 7.13)

$$\text{Correction factor} = \frac{19\,000 \times 0.7}{1000}$$

$$= 13.3$$

The approximate isolux diagram is shown in Fig. 7.14.

With a layout of four uplighters, assuming they are on a spacing of 4 m × 4 m:

the illuminance in the room centre, A, on the isolux diagram, is 4 ×90 lx, i.e. 360 lx,

at B the illuminance is found by calculating its distance from each of the uplighters

$$E_B = 370 + 2 \times 26 + 0$$
$$= 422 \text{ lx}$$

Step 3 – Inter-reflected illuminance (E_R) Step 2 has calculated the illuminance due to first reflection. Light will continue to be reflected and this requires a lumen formula type calculation.

For R_E of 0.5 and RI of 4, the utilisation factor is 0.072 from Table 7.14. Using the lumen formula

$$E_R = \frac{4 \times 19\,000 \times 0.072 \times 0.7}{8 \times 8}$$

$$= 60 \text{ lx}$$

Table 7.14 UF for 2nd and subsequent reflections

Room Reflectances C	W	F	Room Index 0.75	1.00	1.25	1.50	2.00	2.50	3.00	4.00	5.00	00
0.70	0.50	0.20	0.125	0.133	0.135	0.136	0.136	0.134	0.132	0.130	0.128	0.114
	0.30		0.064	0.072	0.077	0.080	0.086	0.090	0.092	0.096	0.099	0.114
	0.10		0.021	0.027	0.031	0.036	0.045	0.052	0.057	0.067	0.073	0.114
0.50	0.50	0.20	0.081	0.083	0.084	0.083	0.080	0.078	0.076	0.072	0.069	0.056
	0.30		0.042	0.45	0.048	0.050	0.051	0.052	0.052	0.052	0.053	0.056
	0.10		0.013	0.016	0.019	0.020	0.024	0.028	0.031	0.035	0.037	0.056

Returning to the direct values, Step 2, the ceiling has an effective reflectance of 0.5 and the Thorn data is based on at least 75% so there is a further scaling if we are to make comparisons. Adjusted direct values plus inter-reflected values give:

$$E_A = 360 \times \frac{0.5}{0.75} + 60$$

$$= 300 \text{ lx}$$

$$E_B = 422 \times \frac{0.5}{0.75} + 60$$

$$= 340 \text{ lx}$$

It is difficult to quote an average but it will be around 310 lx.

Step 4 – check ceiling luminance The CIBSE code recommendation of a limit of 500 cd/m² average and a maximum of 1500 cd/m² apply to the control of discomfort glare. The maximum value can only be checked if the manufacturer provides the intensity distribution of the uplighter. At present manufacturers do not.

The easiest check of average luminance is to consider the ceiling as an 'overcast sky'. In this case the sky luminance, in apostilbs, equals the horizontal ground illuminance. For example:

$$E_H \text{ of } 500 \text{ lx} \equiv L_S \text{ of } 500 \text{ Asb}$$

$$= L_S \frac{500}{\pi} \text{ cd/m}^2$$

$$= 160 \text{ cd/m}^2$$

The scheme will lie well within 500 cd/m² average value.

Comparison of two methods

No direct comparison is possible as Method 1 gives an average illuminance value and is not based on any precise data. However, Table 7.13 (utilisation factors) is for an LOR of 0.75. Scaling up for Method 1:

$$\text{Average illuminance} = 216 \times 0.82/0.75$$
$$= 236 \text{ lx}$$

The two methods do not compare well and method 1 should only be used when precise photometric data is not available.

Emergency lighting

It is an increasing requirement that commercial, industrial, and public buildings are provided with some form of emergency lighting. Requirements vary for different types of buildings and even their geographical location. General guidance is given in BS 5266: 1975 *The emergency lighting of premises.*

The following is a summary of the design requirements. The main classifications are:

1. *Emergency lighting* – lighting provided for use when the normal lighting fails.
2. *Escape lighting* – that part of emergency lighting which is provided to ensure that the means of escape can be safely and effectively used at all material times.
3. *Standby lighting* – that part of emergency lighting which may be provided to enable normal activities to continue.

The basic recommendations are that, during the period of use, the horizontal illuminance measured on the centre line of any escape route *should never fall below 0.2 lx* and in halls and corridors it should be measured at floor level. The emergency system should be in operation *within 5 seconds of the failure* of the normal lighting installation. The ratio of the minimum illuminance to the maximum illuminance along the escape route *should not exceed 1 : 40* and care should be taken to avoid abrupt alternations of excessive dark and light areas on the floor.

There are points in these recommendations that require some elaboration.

1. The illumination level quoted, of 0.2 lx, is roughly equivalent to a moonlit night. It is the intention of the BS that this should be an absolute minimum, i.e. with an aged lamp and battery, soiled luminaire and at the end of the period of duration. Such a situation would suggest a desirable illumination level of between one-tenth and one-fiftieth of the normal lighting level with a minimum value of 1.0 lx.
2. The BS allowance of 5 s before the emergency lighting must be in operation (it can be 15 s for some areas at the local authority's discretion) is based upon practical tests that indicated that after 15 s of darkness people began to move irrationally. Fortunately many of the specialist emergency luminaires are capable of virtually instantaneous operation and the 5 s requirement will be redundant in such cases.

The BS goes on to deal with the problems of specific areas and provides useful comment on the factors that should be considered in the design of an emergency lighting installation. In particular, the merits of the various possible systems are dealt with.

Planning the layouts

Positioning of luminaires

The correct positioning of emergency luminaires is essential in order to provide a system that not only complies with the various legislative requirements but provides a safe and effective way of evacuating a building in the event of a mains failure. Therefore, apart from achieving the minimum light levels, various obstructions, hazards and routes must be covered.

Luminaires and signs should be positioned:

(a) To show exit routes and final exits from premises clearly. Signs should be illuminated;
(b) To ensure exterior areas of final exits are lit to at least the same level as the area immediately inside the exit to enable people to move away from the exit to areas of safety;
(c) Near each intersection of corridors (less than 2 m horizontally);
(d) Near each change of direction (less than 2 m horizontally);
(e) Near each staircase so that each flight of stairs receives direct light (less than 2 m horizontally);
(f) Near any other change of floor level which may constitute a hazard (less than 2 m horizontally);
(g) To illuminate fire alarm call points and fire fighting equipment at all material times;
(h) To ensure normal pedestrian escape routes from covered car parks are illuminated to the same standard as escape routes within buildings;
(i) In plant, switch and control rooms;
(j) Within passenger lift cars, (only self-contained emergency luminaires are suitable for this application);
(k) In toilets exceedings 8 m^2.

In order to meet the illuminance levels manufacturers should provide a table of recommended spacings. A typical example is shown in Table 7.15.

Table 7.15 Spacing for a 4W fluorescent surface mounted luminaire

Ceiling mounting height (m)	*Lighting level directly under luminaires*	*Transverse to wall (m)*	*Transverse spacing (m)*	*Axial spacing (m)*	*Axial to wall (m)*
2.5	1.8 lux	3.8	10.2	9.2	3.4
4.0	0.8 lux	4.0	11.2	10.2	3.6
6.0	0.4 lux	3.2	11.2	10.2	3.0

Spacing Chart for Briklite Luminaire (dimensions in metres).

Undefined Escape Routes

In an open area, such as a hall or shop, it is not possible to define a precise escape route. In this case an average illuminance of 1 lx is required, again with a uniformity of 40 : 1. This may be achieved with separate luminaires or emergency packs in some of the normal luminaires. The lumen method is used to calculate the lamps required for 1 lx and a spacing to height ratio of 2.5 : 1 can give a guide to the minimum number of points.

Calculation:

$$N = \frac{E \times L \times W}{\text{UFO} \times \text{SF} \times \text{ELDL} \times K} \qquad \ldots [7.15]$$

where:

N	=	number of luminaires
UFO	=	utilisation factor at zero reflectance. This is provided by the luminaire manufacturer for a given room index. When using conversion sets in standard mains luminaires the UFO is taken for the mains luminaire
L	=	room length (in metres)
W	=	room width (in metres)
E	=	average illuminance required, i.e. 1.0 lx
SF	=	service correction factor to cover effects of lamp ageing, dust build up, etc., assumed to be 0.8 for normal environments.
ELDL	=	emergency lighting design lumens at nominal volts. For self-contained luminaires the ELDL is given in the specification for each luminaire. For conversion units, multiply the design lumen output of the lamp by the ballast lumen factor.
K	=	a factor to cover the reduction in light output at end of discharge, or 5 seconds after mains failure, whichever is the lowest.

Values of ELDL and K depend on the type of lamp, system and luminaire. It would be misleading to include typical tables in the text. The values should be provided by the supplier of the luminaire. These factors can in some cases effectively reduce the lighting design lumens to less than 10% of the initial lamp lumens.

Maintained operation

This system entails the continuous use of the emergency lighting installation which may be so designed that it appears, under normal conditions, to be part of the conventional lighting installation. It has the advantage that the performance of the emergency lighting is continuously monitored and any lamp or other failure should be instantly noted and dealt with before an emergency situation can arise. As it is usual to supply the emergency system from the normal supply under non-emergency conditions, the fact that the emergency lamps are working satisfactorily does not necessarily mean that the emergency power supply is operational and this must be independently checked.

The alternative power supply can be provided either by a prime mover generator set or by batteries. If a prime mover generator is used it must be capable of being started and run-up to operating speed within 5 s of the failure of the normal supply. To meet such a requirement it may have an automatic starting device which will detect a failure of the normal supply. It requires a high degree of reliability to guarantee that it can be started and run-up within the time specified. As a safeguard, it is possible that a bridging battery can be used to cover the period between failure of the normal supply and the start-up of the generator, which could then be done manually if desired. The BS recommends that a battery capable of at least 1 h duration should be used.

Where batteries alone are used, two methods of connection are possible. The first is the floating maintained system where the lighting load, battery and battery charger are connected in parallel and fed from the normal supply. Should there be a supply failure the battery will continue to feed the emergency lighting installation without the need for any change-over device.

The second is the maintained change-over system. Here the emergency lighting and the battery and charger are separately connected to the normal supply, no load being connected to the battery. In the event of a supply failure an automatic change-over device connects the emergency lighting to the battery. When the normal supply is restored the system reverts to its previous status and the battery is recharged.

Non-maintained operation

This is a system where the lamps are not normally in use but are automatically brought into operation in the event of a normal supply failure. In this case the emergency lamps may either be contained in a separate luminaire or be incorporated in the normal luminaire which may also house the battery, charger and change-over device for the emergency system.

Duration

It is recommended that an absolute minimum duration of 1 h should be provided for even the smallest premises. In some types of premises such as large hotels – particularly those on main thoroughfares where decanting space may be severely limited – it may be desirable to re-occupy the premises immediately the emergency is past or even to delay evacuation if this should be permitted. For these and other reasons a duration of 3 h is recommended for all premises having more than ten bedrooms and more than one floor above or below ground level. Smaller premises should have a duration of 2 h.

As charging periods of 14 h are common for many contained units, problems can arise if the emergency system is exhausted in the early evening and people cannot return until the batteries are recharged or dawn provides the required minimum of 0.2 lx. Such problems are most likely to be solved by discussions with local authority officials and the local fire officer.

8 Energy management and lighting

To obtain the best performance from a lighting installation in terms of energy consumption, and hence cost, it must be efficiently designed, used, and maintained. This chapter introduces a number of topics which must be considered, and while each can produce a saving, *all* the elements should be considered if the saving is to be worthwhile.

Design elements

The various elements can be summarised as follows:

1. The highest utilisation factor compatible with the lighting design requirements.
2. A layout which provides higher illuminance values only where they are required.
3. The highest light loss factor (LLF) compatible with a realistic maintenance programme.
4. The use of appropriate lamps with the maximum luminous efficacy.
5. Switching control system that enables full use to be made of daylight.
6. If the building is suitable, the heat generated by the lighting system to be integrated into other building service requirements.

Use and maintenance elements

7. The lighting only to be used when required.
8. The lamps, luminaires and room surfaces to be cleaned and maintained in accordance with the time schedule selected for the LLF.
9. Advantage to be taken of improvements in lamp efficacies as they occur.
10. The installation should last only as long as it remains efficient i.e. until it can be proved cost effective to replace it with a more up-to-date installation.

Installed lighting loads

The load can be expressed in watts per square metre. There are too many variables for a precise guide on good design to be given, but Table 8.1 gives an indication of values that could be expected for different room indices. The values of watts include the total circuit watts, i.e. the power consumed by the lamp and its control circuit.

Table 8.1 Guide of target wattage for general lighting

Type of interior	*Standard service illuminance (lx)*	*Room index*	*Utilisation factor (guide)*	*Target load for circuit efficacy 60 lm/W*
Heavy industrial	500	5	0.7	15
		2	0.63	16.5
Light industrial	500	5	0.6	17.5
		2	0.53	20.0
Commercial	500	5	0.5	21.0
		2	0.42	25.0

Problem

Referring to Chapter 7, does the lighting scheme designed for the drawing office (see Fig. 7.3) meet the target?

Solution

The loading is twelve 65 W white fluorescent tubes. The control gear will add about 20 per cent to the loading.

$$\therefore \quad \text{Total loading} = 12 \times (65 + (0.2 \times 65))\ \text{W} = 936\ \text{W}$$

The area is 6 m square

$$\therefore \quad \text{Loading} = \frac{936}{6 \times 6}\ \text{W/m}^2 = 26\ \text{W/m}^2$$

This is to produce 750 lx with a circuit efficacy of 5100 lm per 78 W lamp circuit or 65.4 lm/W. The room index is 1.4.

Referring to Table 8.1, the commercial target load for a room index of 2 is 25 W/m². Extrapolating to a room index of 1.4, this becomes 25.5 W/m².

$$\therefore \quad \text{Target load} = 25.5 \times \frac{750}{500} \times \frac{60}{65.4} = 35\ \text{W/m}^2$$

This indicates that the scheme proposed is well within the target load.

Types of layout

Various layouts have already been considered in Chapter 7. The local layout may provide the best opportunity for energy saving as it only provides higher illuminances in the task area with a lower level of general illuminance elsewhere.

An interesting development is the use of 'uplighters' to provide the general illuminance. Such a scheme is shown in Fig. 8.1 and this permits the use of efficient high-pressure lamps such as SON de luxe and MBIF in general office situations. Table 8.2 shows comparative loads for this type of installation and an equivalent general lighting scheme. There are critics of uplighter schemes. The ceilings are too bright for some; the LLF tends to be great unless the maintenance is of a high order; and a single lamp failure

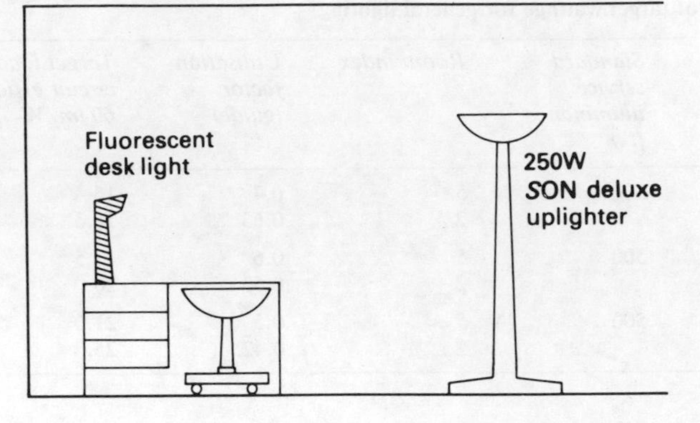

Fig. 8.1 'Uplighter' system for an office

Table 8.2 Comparison of general lighting scheme with uplighter scheme for a general office space

General lighting scheme: Conventional ceiling-mounted fluorescent luminaires to provide 500 lx

Uplighter scheme: 250 W SON deluxe lamps in uplighters to provide 250 lx general lighting; 18 W fluorescent tubes for desk lighting to 500 lx

Scheme	*Loading (W/m^2)*	*Illuminance (lx)*
General	23.5	500
Uplighter/desk light	approx. 8	250/500

causes a fair degree of disruption. However, it is a system which is the subject of considerable interest and research.

Planned maintenance

The LLF has as much effect as the UF on calculating installed load, and if the value is to be any more than a pious hope, it should be linked to a planned maintenance schedule. Unfortunately the engineer who selects the LLF is seldom responsible for establishing a maintenance schedule.

Figure 8.2 illustrates the pattern of light loss due to lamp depreciation and lamp and luminaire soiling. This will vary, depending on the characteristics of the environment, luminaire, and lamp. The figure illustrates that a systematic cleaning and lamp replacement programme will make a significant improvement to the overall performance of the lighting.

Problem

Using the data in Fig. 8.2, what will be the LLF after 9000 h of use if the lamps are replaced after 6000 h and the cleaning interval reduced to 1500 h of use?

Solution

This is illustrated in Fig. 8.3 and is 70 per cent, a considerable improvement on 51 per cent in Fig. 8.2.

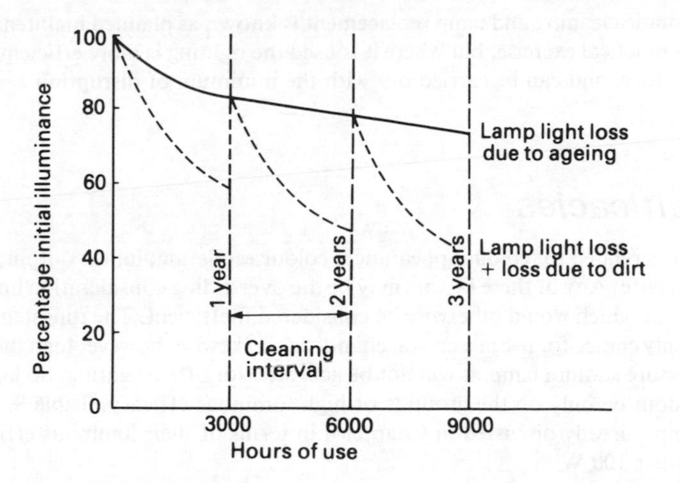

Fig. 8.2 Typical light loss curves due to lamp ageing and dirt collection

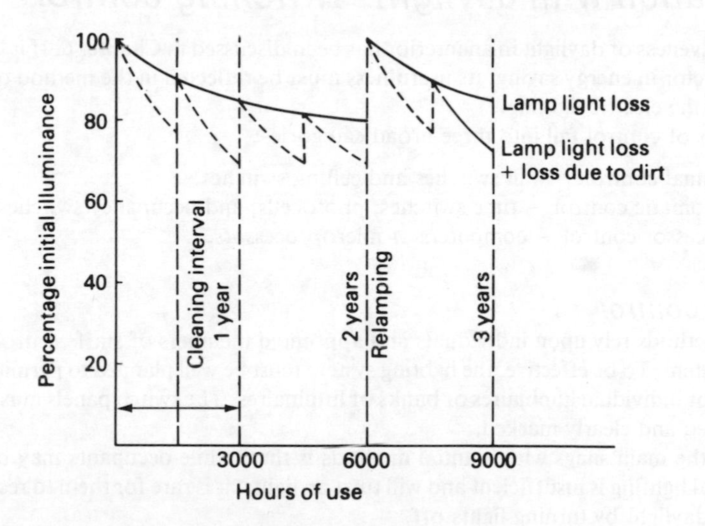

Fig. 8.3 Effect on light loss of 6 monthly cleaning

Table 8.3 Comparative lamp efficacies and costs

Lamp	*GLS*	*TH*	*SOX*	*SON*	*MBF*	*MCF*	*MBI*
Typical lamp luminous efficacy (lm/W)*	12	22	150	100	45	70	75
Approx. lamp cost in £/100 W	0.5	1.6	20	18	6.5	2.8	14
Control gear cost £	—	—	40	25	12	8	25

**These values are based, where possible, on lamps of about 100 W rating*

The systematic cleaning and lamp replacement is known as planned maintenance; it is not always a practical exercise, but where it is used the lighting is more efficient, visually more satisfactory, and can be carried out with the minimum of disruption.

Lamp efficacies

Lamp choice is related to: colour appearance; colour rendering; lumen output; physical size; cost; and life. Any of these factors may be the over-riding consideration limiting the choice to lamps which would otherwise be considered inefficient. The tungsten filament lamp is the only choice for use in a crystal chandelier. Likewise, however high the efficacy of a low-pressure sodium lamp, it will not be accepted for office lighting. So lamp selection can seldom be only on the grounds of high luminous efficacy. Table 8.3 lists the range of lamps already discussed in Chapter 4 in terms of their luminous efficacy and average cost per 100 W.

Integration with daylight: switching control

The effectiveness of daylight in an interior has been discussed in Chapter 6. If it is to be a positive factor in energy saving, its usefulness must be reflected in the method of switch control of the electric lighting.

Methods of control fall into three broad categories:

1. Manual control – wall switches and ceiling switches.
2. Automatic control – time switches, photocells, and occupancy switches.
3. Processor control – computers or microprocessors.

Manual control

Manual methods rely upon individuals and appointed members of staff controlling the lighting system. To be effective, the lighting system must be well planned to permit flexible switching of individual luminaires or banks of luminaires. The switch panels must be sensibly located and clearly marked.

One of the main snags with manual methods is that, while occupants may be aware that natural lighting is insufficient and will turn on lights, it is rare for them to respond to sufficient daylight by turning lights off.

Automatic control

Automatic control in the form of an imposed switch-off (particularly at lunchtime) can be effective, since, if natural lighting is adequate, the luminaires may not be turned back on. A considerable amount of energy can be saved by automatic switching off after working hours. The provision of automatic cleaners' circuits controlling only some of the lighting to provide reduced illuminances can save money.

Automatic control systems can be inexpensive and can switch (or dim) banks of lights. Photocells monitor the level of useful daylight and turn off luminaires or individual lamps in rows adjacent to the windows. Whether or not this is cost effective will depend upon the daylight factor and the proportion of the working year for which the required illuminance is exceeded. Time switches provide a convenient method of ensuring that unwanted lighting is not provided outside working hours.

Occupancy detectors can be used to detect the presence of occupants and control the lights accordingly. These can reply upon acoustic, infra-red, radar, or other methods of detection. A time-lag must normally be built into the system to prevent premature switch-offs.

Processor control

Computer-based or microprocessor-based control systems are becoming increasingly popular, more reliable, and less expensive. These rely upon dedicated computers or processors to control some or all of the building services. An advantage of such an approach is that complex decisions can be taken from moment to moment, based upon the precise state of the building's operation, and that the system is controlled by software. This last feature means that the control programs can be refined and tailored to suit the building and can be amended to suit changed circumstances.

With any control system considerable care must be taken to ensure that acceptable lighting conditions are always provided for the occupants. Safety must always be of paramount importance.

Heat recovery from luminaires

Lamps are comparatively inefficient producers of light, but 100 per cent effective producers of heat. In buildings lit to modern standards, the heat generated by the lighting can make a significant contribution to the heating of the building. It can also be an undesirable heat gain for the refrigeration plant.

Table 8.4 Energy distribution of light sources

Light source	*Fluorescent tube*	*MBF*	*HPS/SON*	*Tungsten*
Convected/conducted (%)	51	30	26	15
Radiant (%)	49	70	74	85
Source temperature (°C)	40	300/800	300/800	2500

Table 8.4 compares the generation of heat by various light sources. In the case of all discharge lamps, a further 10–20 per cent of conducted and convected heat must be added for the control gear. When the lamp wattage is quoted, it does not usually include the control gear. Table 8.5 shows some typical lamp and total circuit wattages.

Table 8.5 Lamp and circuit wattages

Lamp type	*Nominal lamp wattage*	*Approximate total circuit wattage*
Fluorescent	65	80
	85	95
Mercury MBF	125	145
	400	435
Metal Halide MBI	250	291
	400	445
Sodium SON	250	310
	400	450

Effect of luminaire on heat distribution

Lamps radiate a large proportion of their energy, but when they are placed in luminaires a part of that energy is absorbed and converted into conducted and convected energy. The amount depends on the construction and optical system. Figure 8.4 shows a range of typical luminaires and the approximate distribution of energy they emit.

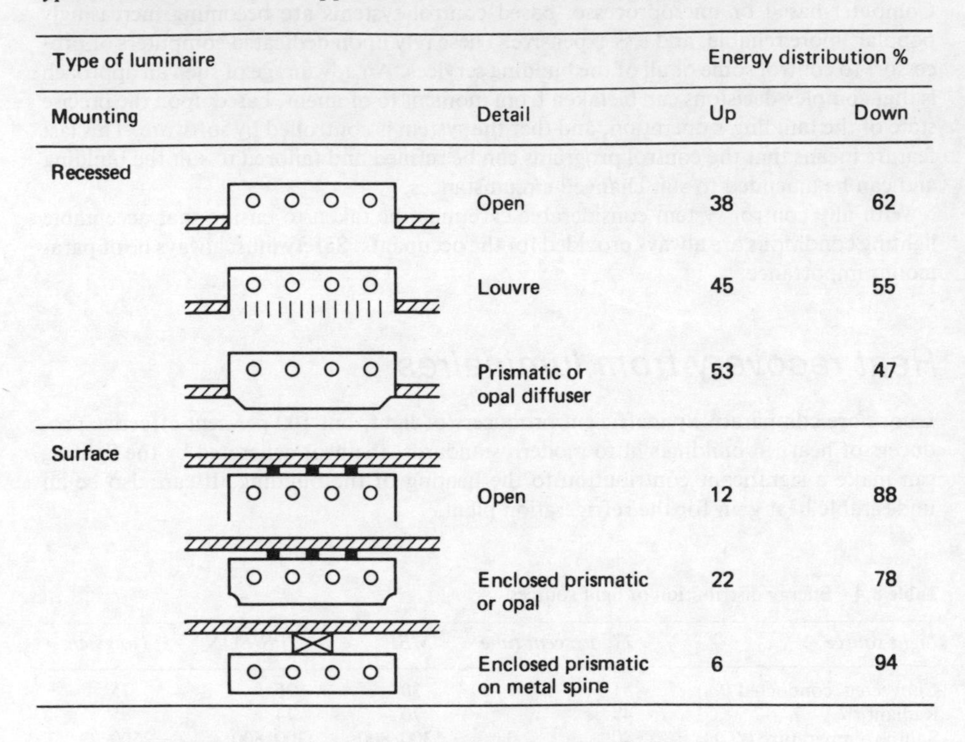

Fig. 8.4 Energy distribution for different types of luminaires

Irradiance from lighting

As already shown, there is a significant amount of radiant power which is a mixture of ultra-violet, light, and infra-red. There is little that can be done to control this heat and people are sensitive to radiant power, in particular on their heads and the backs of their hands.

Table 8.6, which is taken from the CIBSE 1984 Lighting Code, Section B9, shows typical irradiance levels for different types of installation. The values will vary quite considerably, depending on the distance from the luminaires. A person with a bald head will

Table 8.6 Typical irradiance values for different types of fluorescent lighting using white tubes

Mounting	*Luminaire*	*Illuminance (lux)*	*Irradiance (W/m^2)*
Surface	Batten	500	3.6
	Opal diffuser	500	2.6
	Prismatic controller	500	4.5
Recessed	Opal diffuser	500	4.1
	Prismatic controller	500	5.0

be uncomfortably aware of recessed filament lighting in a low ceiling even though the average working plane illuminance may be only 100 lx. The irradiance cannot be eliminated and may even in some situations be desirable, but lamps radiating directly onto people at comparatively short distances should be avoided. An example of a situation needing careful design is the use of track systems in shops where spotlamps may be haphazardly located causing heat discomfort to the shop workers and the customers.

Air-handling luminaires

Luminaires are available which are specifically designed to enable the room air to flow over the lamps, control gear, and metal surfaces, extracting up to 75 per cent of the heat generated by the luminaire.

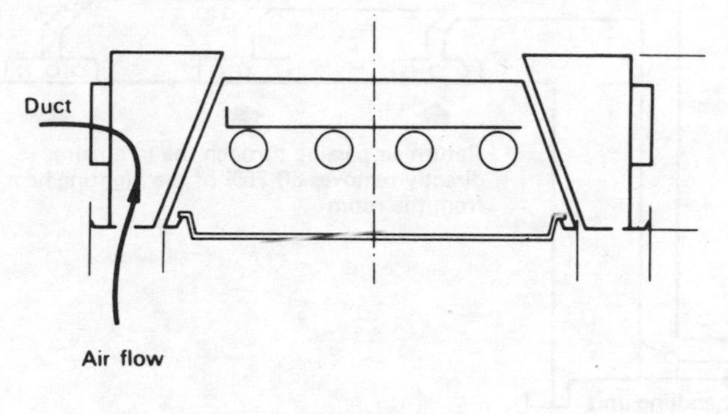

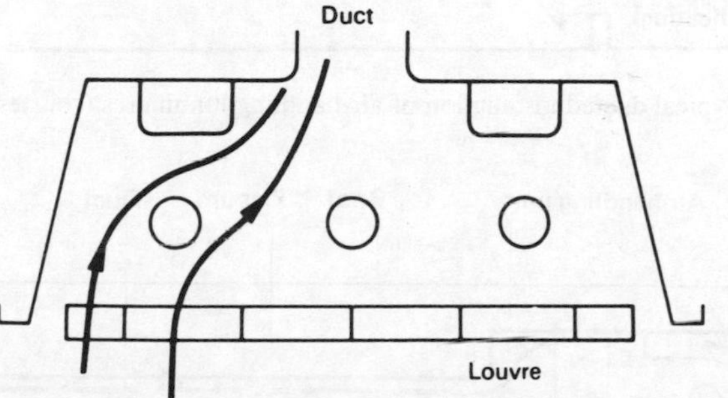

Fig. 8.5 Two forms of air-handling luminaires

Figure 8.5 shows typical designs, but there are many variations depending on the method of integrating the air-handling components with the general mechanical services. The advantages of using such a system include:

1. A more economic use of the energy within a building. For instance refrigeration load is not required to counteract heat generated by the lamps.
2. Fluorescent lamps run at a more efficient temperature. The effect of temperature on light output has already been discussed in Chapter 4 and improvements of up to 10 per cent can normally be expected.
3. The visual environment is improved by the integration of ceiling services. Often where lighting and air handling are designed as separate services, their layouts conflict giving an untidy effect on the ceiling.

This form of heat recovery is usually feasible where the gross heat gain from the lighting, occupants, and equipment is comparatively large and the interior is well insulated thermally. It may suit a deep office, but could be quite unsuitable for a shallow building with a high percentage of the exterior wall glazed.

Method of installation

Air can be extracted from the room through the luminaire. This is shown in Fig. 8.6 and the system would be fully ducted. The ductwork for the return air can be reduced by exhausting through the luminaire into the plenum, as shown in Fig. 8.7. For this to work, the whole of the ceiling system must be reasonably airtight. Leakage of air through the ceiling can unbalance the system, cause ceiling tile pattern staining, and reduce the heat transfer efficiency of the luminaire.

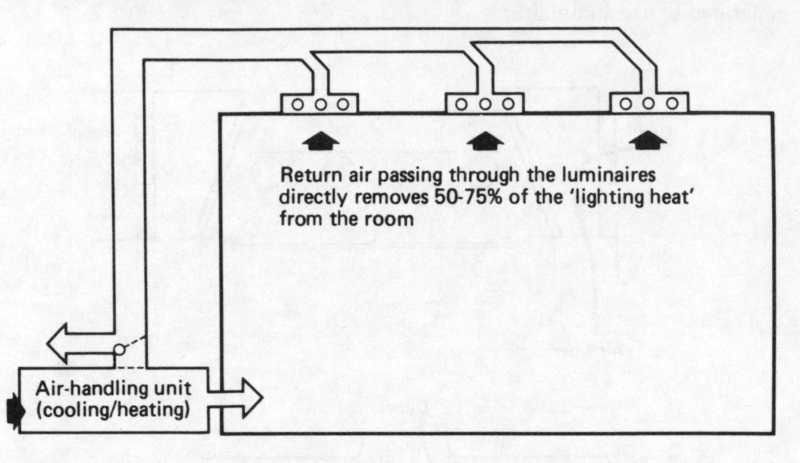

Fig. 8.6 Typical ducted installation of air-handling luminaires (courtesy The Electricity Council)

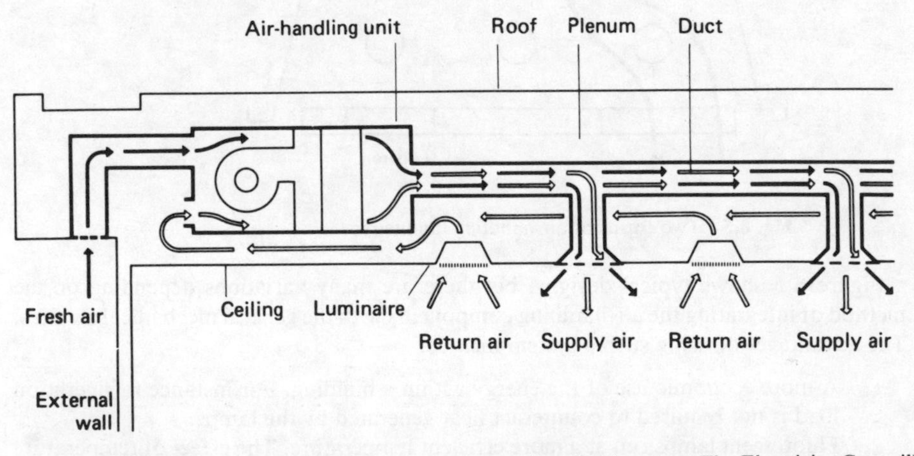

Fig. 8.7 Use of the plenum as an exhaust chamber (courtesy The Electricity Council)

Performance of luminaires

Apart from the usual photometric data, a manufacturer of air-handling luminaires must supply sufficient data for the heating and ventilation calculations.

To obtain these data the luminaire must be tested in a calorimeter and the following is normally required:

(1) total heat flow to plenum;
(2) rate of heat input (electrical load);
(3) exhaust air temperature;
(4) room air temperature;
(5) relative light output;
(6) air flow rate.

The construction of a standard calorimeter is outlined in Section B9 of the CIBS 1984 Lighting Code together with the test procedure. Test data is normally presented in graphical form and a typical set of results is shown in Figure 8.8.

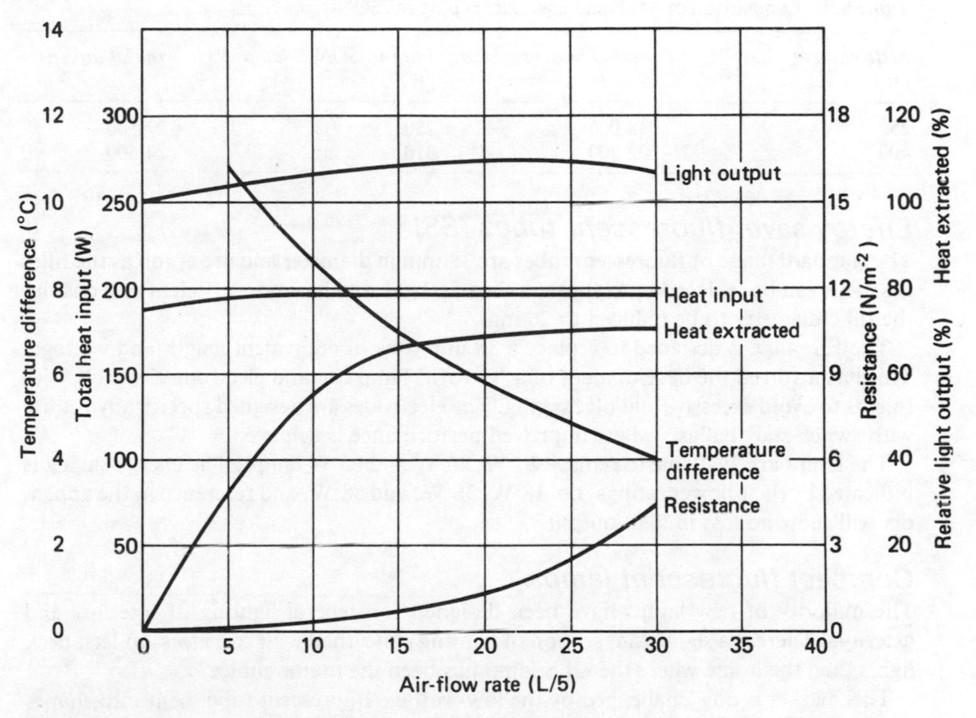

Fig. 8.8 Typical data sheet for an air-handling luminaire (courtesy Moorlite Electrical Ltd.)

Energy-saving developments

The lighting industry has produced many ideas to make effective savings of energy. Some have been little more than sales gimmicks; others have been of value. The aim is normally to achieve the same or better illuminance from an existing installation by using different lamps. The following are some of the more successful ideas.

Blended lamps

If GLS lamps are in use, improved efficiency normally requires replacement with discharge lamps. These need control gear, which implies new luminaires. Blended lamps are mercury discharge lamps incorporating a series resistance in the form of a tungsten filament. This results in a lamp of around 20 lm/W with a good standard of colour rendering. It is a compromise in that it is not as efficient as a standard MBF lamp but it is a method of upgrading a GLS installation.

Plug-in lamps

High-pressure mercury lamps have been available since the mid-1930s. They have proved an economic and reliable light source but there are many installations using 250 W or 400 W mercury lamps which could be significantly upgraded in efficiency and light output. Table 8.7 compares the performance of the mercury lamp and its equivalent plug-in high-pressure sodium lamp. Savings of 15 per cent in power consumption and up to 50 per cent higher light output can be achieved by simply replacing the mercury lamp with the equivalent plug-in sodium lamp with no change to the electrical control circuit.

Table 8.7 Comparison of MBF and equivalent 'plug-in' SON

MBF mercury lamp (W)	*Initial lumens*	*Plug-in SON-E lamp (W)*	*Initial lumens*
250	13 000	220	22 500
400	22 700	310	24 500

Energy-saver fluorescent tubes (ES)

The standard range of fluorescent tubes are 38 mm in diameter and use argon as the filler gas. This can be replaced by krypton, a 'heavier' gas which is more efficient and enables the tube diameter to be reduced to 26 mm.

The ES range is designed to replace a 38 mm tube of equivalent length and wattage. This has involved the development of a 'retrofit' lamp cap and electrode assembly with shields to avoid excessive and blackening. The electrodes are designed specifically for use with switch-start ballasts where improved performance is achieved.

The lamps are available to replace 20 W, 40 W and 65 W lamps. The energy saving is indicated by their power ratings, i.e. 18 W, 36 W, and 58 W, and reference to the appendix will show no loss in light output.

Compact fluorescent lamps

The majority of new lamps have been designed for general lighting of interiors and exteriors. There is a significant section of lighting concerned with corridors, toilets, desk lights, and the home where the GLS lamp has been the major choice.

This market is now challenged by the low-wattage fluorescent tube, bent into shapes that enables it to fit into small luminaires. Figures 8.9 and 8.10 show two such lamps. The SL lamp can be used as direct replacement to a GLS lamp. The 'D' lamp is used in luminaires specifically designed for it. Table 8.8 compares their performance with GLS lamps. However, one word of caution. Power factor correction is not normally required for lamps below 30 W and the saving in volt-amps may not be as satisfactory as the saving in watts.

Table 8.8 Comparison of GLS lamp with compact fluorescent lamps

GLS (W)	*Initial lumens*	*2D (W)*	*Initial lumens*	*SL (W)*	*Initial lumens*
60	610				
100	1230	16	1050	9	450
150	2060	28	2050	13	650
				18	900
				25	1200

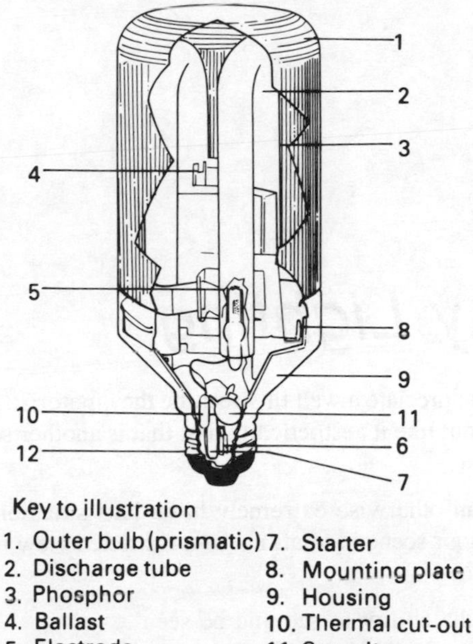

Key to illustration

1. Outer bulb (prismatic)
2. Discharge tube
3. Phosphor
4. Ballast
5. Electrode
6. Bi-metallic strip
7. Starter
8. Mounting plate
9. Housing
10. Thermal cut-out
11. Capacitor
12. Lamp cap

Fig. 8.9 SL compact fluorescent lamp (courtesy Philips Electronic Ltd)

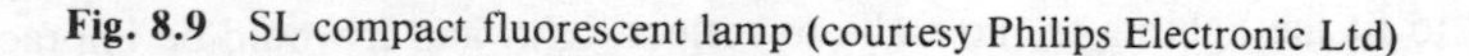

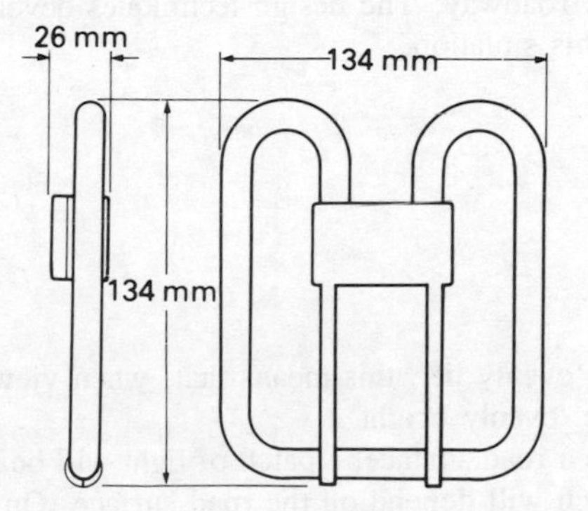

Fig. 8.10 A 2D 16W compact fluorescent lamp (courtesy Thorn EMI Lighting Ltd)

High frequency circuits — see p. 67.

Conclusion

As long as energy costs rise, energy saving schemes will develop. It would be pleasing to think that energy-saving schemes are introduced to save energy. The reality is that they are only introduced to save cost, and if this is not achieved, the ideas are abandoned. Costing a scheme is a complex process involving energy tariff, interest rates, and taxation methods. The best form of economy is to design an efficient and effective scheme and to maintain it properly.

9 Roadway Lighting

There are few people who do not appreciate a well lit street, be they motorist, passenger, cyclist, or pedestrian. They may not like it aesthetically, but that is another story. What do they appreciate?

1. The degree of safety in an otherwise extremely hazardous situation;
2. The ability to see the outdoor scene in total, not as a restricted view in the beam of a car headlight.

These could be summarised as 'the ability to see and be seen'.

How does one define 'a well lit street'? It would have to appear evenly illuminated, with no obvious dark spaces, and a minimum of glare.

How is this achieved? Not in the same way as a well lit interior. There is no convenient 'ceiling' to fit the lighting to, no walls to reflect light, and it would be impractical to maintain even illumination along the roadway. The design techniques developed for interior lighting are of little use in this situation.

Silhouette vision

When a road is referred to as being 'evenly lit', this means that, when viewed from a car, the road surface appears to be 'evenly bright'.

If a lamp is hung some 10 m above a road surface, a patch of light will be reflected from the road. The shape of this patch will depend on the road surface. On surfaces of very fine texture which take a noticeable polish, such as asphalt, the patch will be long, extending even to the feet of the observer. On the more usual rough roads, the patch will extend across the road rather than down it. Hardly any bright area will be seen on the far side of the post supporting the lamp (Fig. 9.1). The patches are not so well defined as shown in the sketch, but can nevertheless be observed quite distinctly.

If a succession of lanterns along the length of the road is so arranged that the bright patches merge to cover the road area, objects on the road will be seen as dark silhouettes against the bright surface (Fig. 9.2). This is the principle upon which most street lighting is based, since it proves more economic to produce silhouettes than it would be to make objects light and the road surface dark.

By day the illuminance on the road surface is approximately constant along the length. By night the relation is quite different. The maximum illuminance on the road will be

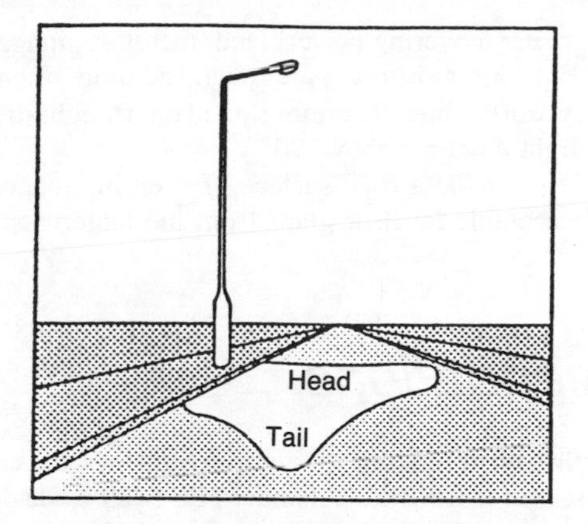

Fig. 9.1 Luminance patch from a single lamp

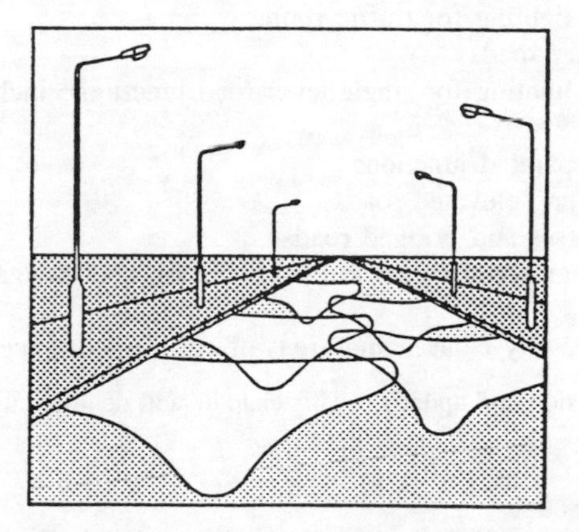

Fig. 9.2 Luminance patches merging to give even lighting

immediately below the hypothetical bare source and will tail off up and down the road dropping to 1/10 the maximum at about 12 m and to a negligible value at twice this distance. The luminance of the surface, however, to an observer some 70 m down the road will have its maximum at about 12 m from the foot of the column, falling off steadily closer to him.

The long bright patch is usually described as 'T-shaped', having a 'head' and a 'tail'. The head is formed by light emitted from the lantern over the range of angles from vertically downward up to about 60° (i.e. 30° below the horizontal). Light above the 60° angle forms the tail.

Glare

One objection to lighting a main road with rows of powerful bare lamps would be the glare. Examination shows that much of this arises from light emitted between 70° and the horizontal, i.e. the light which produces the long tail.

If lighting highly polished roads or permanently wet ones, it may be worthwhile to accept this glare in return for the greater 'covering-power' (and, therefore, longer spacing and cheapness) of the bare lamp. But since most roads are rough and more often dry than wet, this exchange is not generally worthwhile. Therefore, the lantern light distribution is so designed that only a little light emerges above 70°.

The aims of good lighting are to provide a road surface of even luminance as seen by the road user without an unacceptable level of glare from the lanterns.

Design recommendations

A very complex design exercise has been reduced to fairly straightforward calculation and design processes. These are set out in the British Standard BS 5489 Code of practice for road lighting.

Part 1: Guide to the general principles
Part 2: Code of practice for lighting for traffic routes
Part 3: Lighting for subsidiary roads
Part 4: Code of practice for lighting for single-level road junctions, including roundabouts
Part 5: Lighting for grade-separated junctions
Part 6: Lighting for bridges and elevated roads
Part 7: Lighting for underpasses and bridged roads
Part 8: Lighting for roads near aerodromes, railways, docks, and navigable waterways
Part 9: Lighting for town and city centres and areas of civic importance.

These are under continuous revision and updating. This chapter will deal mainly with the contents of Parts 1, 2, 3 and 4.

Part 1: Guide to the general principles

This is a relatively short section. Its purpose is to state the principles on which Parts 2 to 9 are based. It includes definitions, road classifications, daytime appearance, the visual scene and task at night, the intentions of road lighting, the hazards of columns, the law, maintenance.

In understanding the design process the following is of help.

Definitions

Arrangement. The pattern according to which lanterns are sited in plan, e.g. staggered, opposite, single side and twin central.

Average luminance (of the road surface). The average luminance over a defined area of the road surface viewed from a specified observer position.

Design attitude (of a lantern). The disposition of a lantern in space, usually indicated by a diagram or by reference to a datum axis such as the spigot entry.

Design spacing. The required spacing between lanterns, calculated as specified in the various Parts of BS5489, for a straight and level section of the particular type of road.

Effective width (W_E). A notional distance, related to the width of the carriageway, the lantern overhang and the arrangement, used to simplify the design tables.

Equivalent veiling luminance (Lv). Luminance that is, in effect, super-imposed on the visual scene as a result of bright areas in the field of view.

Geometry (of a lighting system). The interrelated linear dimensions and characteristics of the system, i.e. spacing, mounting height, effective width, overhang and arrangement.

Longitudinal uniformity ratio (of luminance) (U_L). The ratio of the minimum to the maximum luminance along a longitudinal line through the observed position on the carriageway.

Mounting height. The nominal vertical distance between the photometric centre of a lantern and the surface of the carriageway.

Overall uniformity ratio (of luminance) (U_O). The ratio of the minimum to the average luminance over a defined area of the road surface viewed from a specified observer position.

Overhang (A). The distance measured horizontally between the photometric centre of a lantern and the adjacent edge of a carriageway. The distance is taken to be positive if the lanterns are in front of the edge and negative if they are behind the edge.

Set back. The shortest distance from the forward face of a column to the edge of a carriageway.

Spacing. The distance, measured parallel to the centreline of the carriageway, between successive lanterns in an installation.

Surround ratio (S_R). The ratio of the average illuminance on a strip 5 m in width beside the carriageway to the average illuminance on the adjacent 5 m strip of carriageway.

Threshold increment (TI). A notional measure of the effect of disability glare from street lanterns.

Maintenance factor (MF). A factor which allows for the depreciation in the photometric performance of a lantern from its state when new to its worst acceptable state in service.

Spacing index (S_I). The product of the calculated average road luminance (in cd/m^2) produced by an installation of lanterns in their clean state and the spacing (in m) divided by the lamp flux (in klm).

Width of carriageway (W_K). The distance between kerb lines, measured at right angles to the length of the carriageway.

Road classification

The full classification is as described in Parts 2–9. The Parts of main concern in this chapter are:

(a) Lighting for traffic routes (BS5489: Part 2).
(b) Lighting for subsidiary roads (BS5489: Part 3).
(c) Lighting for single-level road junctions including roundabouts (BS5489: Part 4).

Part 2: Code of practice for lighting for traffic routes

This is the 'meat' of BS5489 as it contains the necessary instructions on the design of lighting for all-purpose traffic routes up to 15 m in width for single carriageways, and up to 2 × 11 m in width for dual carriageways. It does *not* deal with motorways. The following sections deal with the calculations but a copy of the BS5489 is essential for the complete design process.

Mounting Height (H)

In general, increasing *H* will enable wider spacing to columns. This may be of help on long open stretches of road, but not so in urban situations with many junctions and intersections.

The choice of *H* is 8 m, 10 m, and 12 m. A height of 10 m is the recommendation for the majority of roads, with 12 m for wide or heavily used roads such as a major route between two towns.

Control of glare

Disability is the criterion (rather than discomfort). It lends itself better to numerical treatment and it is considered that limiting disability glare will also limit discomfort glare.

Glare is a function of the source luminance and the luminance of the background. An object that is just visible (that is at the threshold of visibility) when there is no disability glare will, in the presence of disability glare, merge into the background. The percentage by which the background luminance has to be increased to render the object just visible again is known as the threshold increment (TI). This provides a notional measure of disability glare.

The value of TI depends on the light distribution from the lantern between 70° and 90° in elevation in the vertical plane at which the lantern is observed. It also depends on the road luminance, the layout of the lanterns, the mounting height and the observer position.

A value of TI not exceeding 15 per cent is recommended where it is necessary to minimise glare, e.g. on high speed roads. It is of particular advantage in rural areas, where the absence of a reflecting background, such as buildings, may increase the effect of glare. In all other installations the value of the TI should not exceed 30 per cent.

Manufacturers normally classify their lanterns LTI (low threshold increment) to meet the 15 per cent requirements or MTI (moderate threshold increment) to meet the 30 per cent value. The British Standard gives a method for working out the TI but suggests this will only be needed on long straight roads where a large number of lanterns are visible.

Surround Ratio

This has already been defined (p. 163) and it is a method of ensuring that there is adequate visibility for pedestrians and that vehicle drivers can see people and objects to the side of the carriageway. An S_R value of not less than 0.5 is suggested and this can be calculated from the isolux diagram.

Luminance of road surfaces

The aim of good lighting is to achieve a road surface of even brightness. The eveness is expressed in terms of U_O (overall uniformity ratio – see definitions). This will depend on the type of road surface, lantern, and the geometry of the layout. Another term U_L (longitudinal uniformity ratio) is also used. U_O is of such a value that no part of the road is too dark and U_L of a value so that the appearance is not too 'patchy'.

Table 9.1 Recommended values of luminance and uniformity ratio (from BS5489 Pt 2)

Category	*Average luminance* $\bar{L}$ *(in cd/m²)*	*Overall uniformity ratio* U_o	*Longitudinal uniformity ratio* U_L	*Examples*
1	1.5	0.4	0.7	High speed road. Dual carriageway roads
2	1.0	0.4	0.5	Important rural and urban traffic routes Radial roads District distributor roads
3	0.5	0.4	0.5	Connecting, less important roads Local distributor roads Residential area major access roads

The average level of brightness is expressed in luminance (cd/m^2). Table 9.1 shows the BS recommendations.

To achieve the values in Table 9.1 the layout of columns and lanterns (the geometry) is planned with the help of data and tables contained in Part 2.

Figures 9.3, 9.4, 9.5 and 9.6 show all the important dimensions and what is meant by a 'staggered', 'opposite', 'single-sided' and 'central' layout.

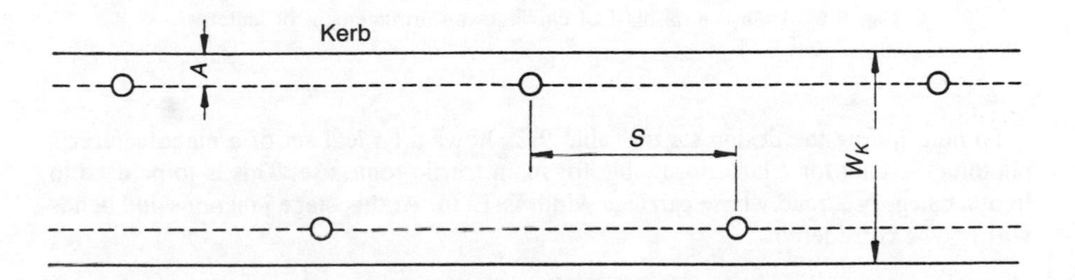

NOTE. Effective width (W_E) is $W_K - 2A$ for lanterns over the carriageway and $W_K + 2A$ for lanterns behind the edge of the carriageway.

Fig. 9.3 Staggered arrangement of lanterns

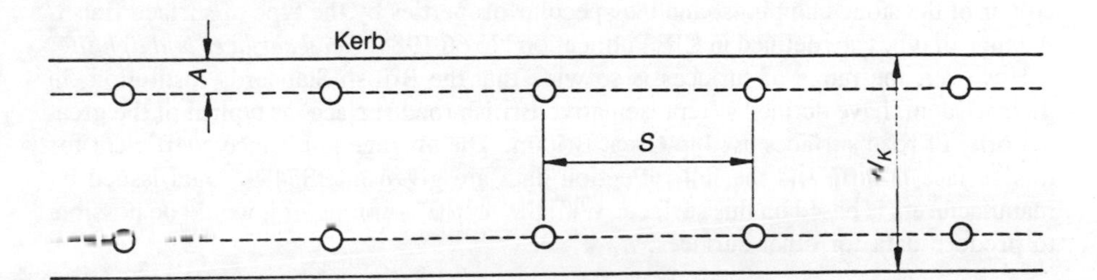

NOTE. Effective width (W_E) is $W_K - 2A$ for lanterns over the carriageway and $W_K + 2A$ for lanterns behind the edge of the carriageway.

Fig. 9.4 Opposite arrangement of lanterns

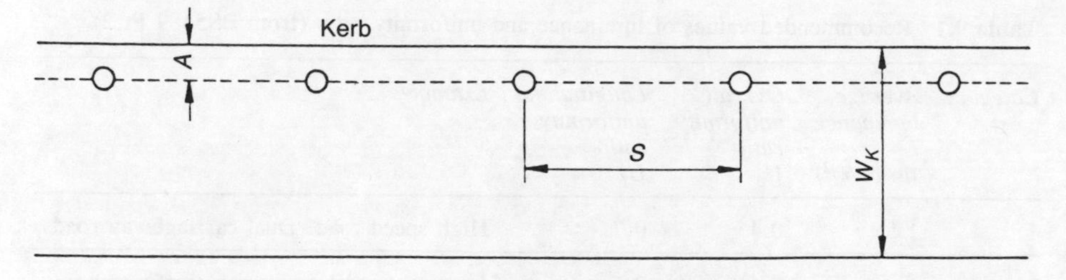

NOTE. Effective width (W_E) is $W_K - A$ for lanterns over the carriageway and $W_K + A$ for lanterns behind the edge of the carriageway.

Fig. 9.5 Single side arrangement of lanterns

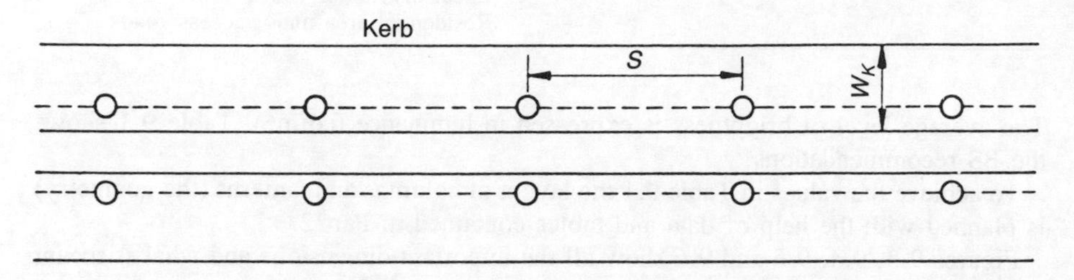

NOTE. Effective width (W_E) is W_K.

Fig. 9.6 Twin central on dual carriageway arrangement of lanterns

To help follow the design steps, Table 9.2 shows a typical set of a manufacturer's photometric data for a lantern suitable for main traffic route use. This is to be used to light a category 2 road whose carriage width is 12 m. At this stage junctions and bends will not be considered.

Road surface

Luminance values of the road surface depend on the light received and the reflectance properties of the road surface. The latter comprises both the total reflectance and the specular reflectance at various angles. The total reflectance is mainly determined by the colour of the stone chippings, and the specular properties by the type of surface finish. A range of types are defined in CIE Publication No 66 1984: *Road surfaces and lighting*.

However, the range of surfaces is so wide that the British Standards Institution, in their wisdom, have defined a 'representative British road surface' as typical of the great majority of road surfaces used in Great Britain. The average luminance coefficient for this surface is 0.07 and the full reflection data are given in BS5489. Data issued by manufacturers is based on this surface. With the help of a computer it would be possible to produce data for other surfaces.

Geometry of the layout

Assume a staggered layout with a 2 m overhang (A). Also assume that MTI will meet the glare control.

Table 9.2 Design table: standard presentation

Lantern Lamp: 150 W SON/T Design attitude: Spigot entry elevated 5° Mounting height: 10 m Lantern classification MTI

Staggered

$S(m)$	W_E 6 m S_I3.78 S_R0.83			W_E 7 m S_I3.58 S_R0.84			W_E 8 m S_I3.38 S_R0.85			W_E 9 m S_I3.19 S_R0.85			W_E 10 m S_I3.02 S_R0.86		
	U_O	U_L	V_F	U_O	U_L	V_F	U_O	U_L	V_F	U_O	U_L	V_F	U_O	U_L	V_F
20	0.68	0.81	29	0.68	0.80	29	0.67	0.81	27	0.66	0.83	27	0.66	0.83	26
22	0.66	0.81	27	0.65	0.80	26	0.65	0.80	26	0.65	0.82	25	0.65	0.82	25
24	0.65	0.82	25	0.64	0.81	24	0.65	0.81	24	0.65	0.81	24	0.66	0.84	23
26	0.62	0.73	24	0.62	0.73	23	0.62	0.72	23	0.63	0.73	23	0.63	0.74	22
28	0.63	0.75	22	0.62	0.74	22	0.63	0.71	22	0.64	0.70	22	0.63	0.70	22
30	0.63	0.73	21	0.63	0.73	21	0.63	0.71	21	0.61	0.71	20	0.59	0.71	20
32	0.63	0.67	21	0.63	0.65	20	0.61	0.63	20	0.58	0.62	19	0.57	0.62	19
34	0.63	0.67	19	0.60	0.63	19	0.58	0.61	19	0.56	0.61	19	0.54	0.58	18
36	0.61	0.67	19	0.56	0.63	18	0.54	0.58	18	0.50	0.56	18	0.49	0.55	18
38	0.59	0.64	18	0.58	0.61	18	0.54	0.58	18	0.49	0.56	17	0.47	0.53	17
40	0.55	0.58	17	0.53	0.56	17	0.51	0.54	17	0.50	0.52	16	0.48	0.50	16
42	0.54	0.57	17	0.52	0.54	17	0.51	0.52	16	0.50	0.50	16	0.47	0.48	16
44	0.56	0.56	16	0.53	0.53	16	0.51	0.51	16	0.49	0.49	16	0.46	0.48	15
46	0.55	0.56	16	0.52	0.52	16	0.50	0.50	15	0.48	0.49	15	0.46	0.47	15
48	0.53	0.55	16	0.50	0.51	15	0.47	0.49	15	0.46	0.47	15	0.45	0.45	14
50	0.52	0.50	15	0.49	0.46	15	0.46	0.44	14	0.45	0.42	14	0.44	0.40	14

$S(m)$	W_E 11 m S_I2.84 S_R0.88			W_E 12 m S_I2.71 S_R0.90			W_E 13 m S_I2.62 S_R0.92			W_E 14 m S_I2.48 S_R0.95			W_E 15 m S_I S_R			W_E S_I S_R		
	U_O	U_L	V_F	U_O	U_L	V_F	U_O	U_L	V_F	U_O	U_L	V_F	U_O	U_L	V_F	U_O	U_L	V_F
20	0.65	0.82	26	0.65	0.82	25	0.65	0.82	24	0.64	0.82	24						
22	0.65	0.82	24	0.65	0.81	22	0.65	0.81	23	0.65	0.81	22						
24	0.66	0.83	23	0.66	0.82	22	0.65	0.83	21	0.66	0.83	21						
26	0.63	0.74	21	0.62	0.74	21	0.61	0.74	21	0.61	0.72	18						
28	0.62	0.70	21	0.62	0.70	20	0.60	0.69	19	0.60	0.69	18						
30	0.58	0.70	19	0.58	0.69	19	0.58	0.69	18	0.57	0.68	18						
32	0.55	0.62	19	0.53	0.62	18	0.51	0.61	18	0.50	0.62	17						
34	0.52	0.58	18	0.51	0.57	17	0.48	0.56	17	0.48	0.56	16						
36	0.46	0.53	17	0.45	0.54	16	0.44	0.54	16	0.44	0.53	16						
38	0.45	0.51	16	0.43	0.50	16	0.41	0.49	16	0.40	0.49	16						
40	0.44	0.49	16	0.42	0.48	16	0.40	0.47	15	0.40	0.45	15						
42	0.45	0.47	16	0.43	0.46	15	0.42	0.45	15	0.41	0.44	15						
44	0.44	0.46	15	0.43	0.45	15	0.42	0.44	14	0.40	0.43	14						
46	0.44	0.47	14	0.43	0.45	14	0.40	0.44	14	0.40	0.44	14						
48	0.43	0.44	14	0.40	0.43	14	0.37	0.42	14	0.37	0.42	13						
50	0.42	0.39	14	0.39	0.38	13	0.36	0.38	13	0.34	0.38	13						

Next refer to Table 9.2 to check that S_R meets the requirements of 'not less than 0.5'. The value in the table will be for W_E of 8 m (Fig. 9.3) and this is 0.85 which is satisfactory.

Next the spacing index S_I is seen to be 3.38. This index is defined as:

$$S_I = \frac{S\bar{L}}{\phi\,\text{MF}} \qquad \ldots [9.1]$$

where S is spacing in metres
$\bar{L}$ is the average road luminance (in cd/m^2)
ϕ is the lighting design lumens (in klm)
MF is the maintenance factor (see Table 9.3).

This formula can be rearranged to give an actual spacing S:

$$S = \frac{S_I\,\phi\,\text{MF}}{\bar{L}}$$

If the lamp is 150 W SON/T then ϕ is 15.5 klm. MF from Table 9.3 for a cleaning interval 12 months, protection of IP 54, and medium pollution, is 0.88. The value of L for category 2 is 1 cd/m^2 (Table 9.1).

$$\therefore \quad S = \frac{3.38 \times 15.5 \times 0.88}{1}$$

$$= 46.1 \text{ m.}$$

Referring back to Table 9.2 for S of 46 m, W_E of 8 m

$U_O = 0.50$
$U_L = 0.50$
$V_F = 15$

Referring back to Table 9.1, U_O and U_L meet the uniformity requirements. V_F is the 'veil factor' which can be used to calculate the threshold increment TI. However it is normally reasonable to accept the manufacturers' classification.

The result of these calculations indicate a satisfactory layout. There are a number of permutations of mounting height (H), lantern type (MTI or LTI), lamp flux (ϕ) and layout. The final scheme is a result of cost, the designer's experience, and aesthetics.

Table 9.3 Maintenance factors

Cleaning interval	*Degree of protection of lamp housing*					
	IP23 minimum			*IP54 minimum*		
	Pollution category			*Pollution category*		
	High	*Medium*	*Low*	*High*	*Medium*	*Low*
months						
6	0.61	0.69	0.96	0.91	0.92	0.96
12	0.53	0.62	0.94	0.86	0.88	0.94
18	0.48	0.58	0.92	0.83	0.85	0.92
24	0.45	0.56	0.91	0.81	0.83	0.91
36	0.42	0.53	0.90	0.79	0.82	0.90

Spacing on bends

The pattern of luminance patches interlock on a straight stretch of road. Even a gentle bend in plan, when seen by the driver cuts off all but the closest patches. To maintain even luminance it may be necessary to reduce the spacing *S* and only site the columns on the outside of the bends.

To help design the layout, a transparent template is made up as shown in Fig. 9.7. This shows the isoluminance contours of 12.5 per cent max (A) and 25 per cent (C). It is drawn to the same scale as the road plan, e.g. 1 : 500 and the point P represents a driver 90 m from the lantern. The template will only apply to a specific *H*, in this case 10 m.

Use of the template is shown in Fig. 9.8. It indicates the patch of luminance that the driver will see from various points around the bend. The shaded area indicates a possible dark patch and the spacing *S* needs to be reduced. The patterns should be checked for both directions (except for a dual carriageway).

Curve B is the 12.5 per cent contour for 2 lanterns back-to-back.

Road junctions

There are a wide variety of junctions and these are dealt with in BS5489 Part 4. Junctions are the most likely places for accidents and the lighting must:

(a) warn the driver he is approaching a junction,
(b) reveal any traffic emerging from minor roads,
(c) not provide additional hazard of columns sited at the junction blocking the view.

Figures 9.9, 9.10, 9.11 and 9.12 show typical layouts for:

A tee-junction – Fig. 9.9
A crossroad – Fig. 9.10
A mini-roundabout – Fig. 9.11
A roundabout – Fig. 9.12

The principle is to maintain even luminance, site a column beyond an intersection (no more than $\frac{1}{3}$ *S*) and, where possible, keep to the outside of bends. When approaching a major road from a minor one try to site a column on the far side of the intersection. It all makes sense if one can anticipate what drivers need to see, whether they have to halt or be aware of any emerging traffic.

Pedestrian crossings

Similar principles apply to pedestrian crossing and Fig. 9.13 shows layouts for staggered and opposite layouts. The layout becomes more complex when crossings are at road junctions. It is to be hoped that the junction layout is satisfactory for the crossing.

Part 3: Code of practice for subsidiary roads and associated pedestrian areas

Main traffic routes consider the needs of the driver, and everything else, pedestrians included, is a hazard to be seen and avoided. Many roads with lower traffic speeds and density are more social areas where the aims are not so heavily driver orientated. The pedestrian is more concerned with illumination and seeing in a similar fashion to an interior situation.

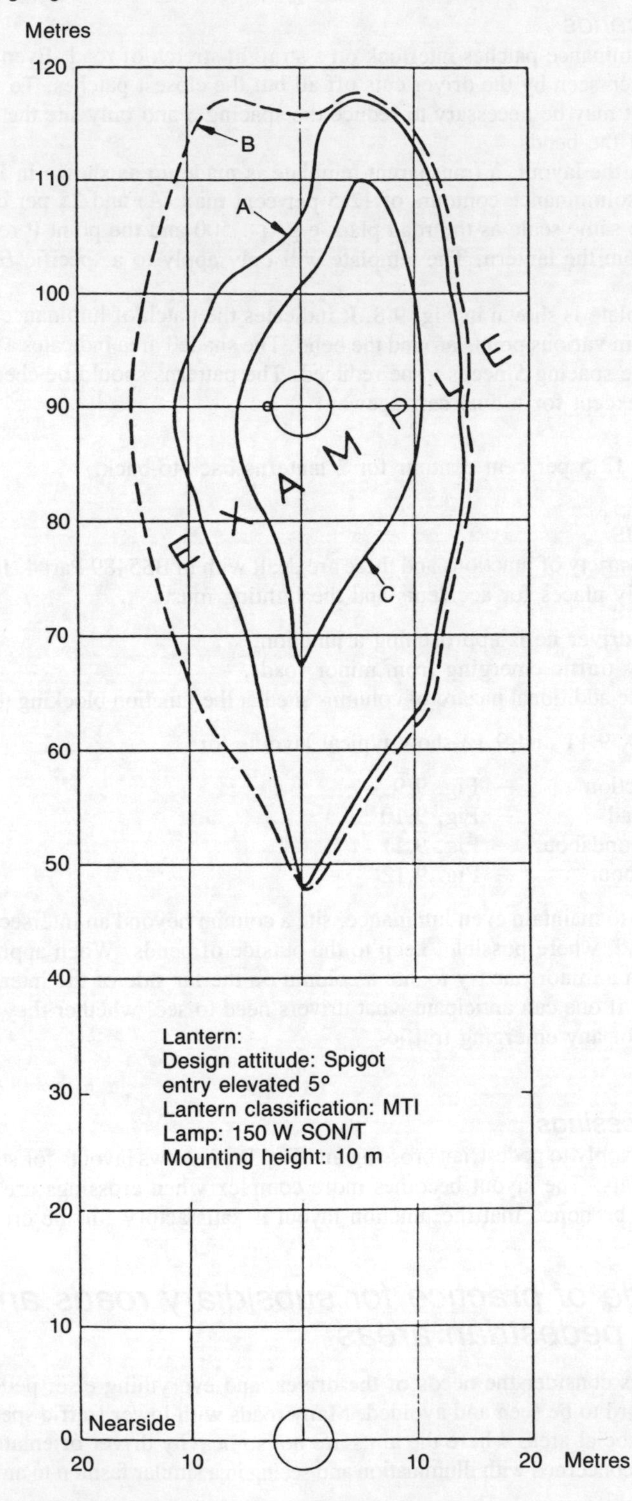

Fig. 9.7 Isoluminance templates

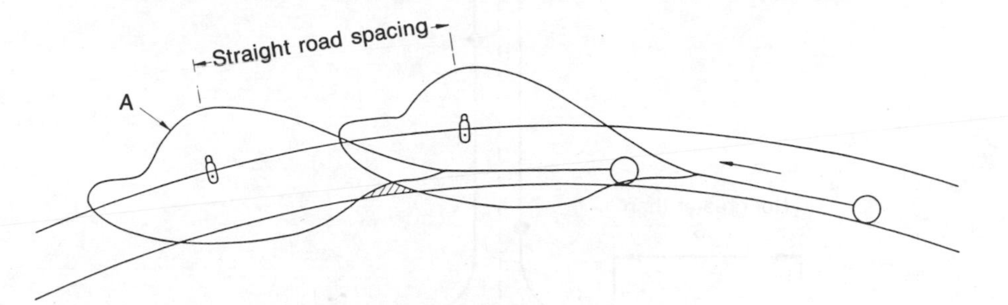

Fig. 9.8 Inadequate coverage of road

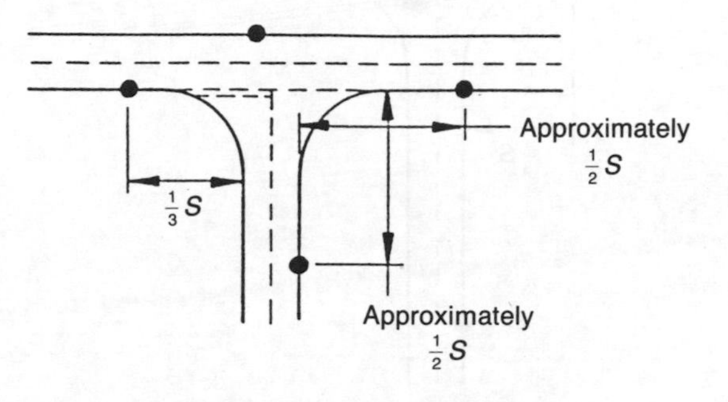

S is the design spacing for the major road.

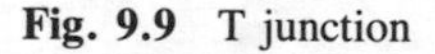

Fig. 9.9 T junction

The recommendations are given in the form of average and minimum horizontal illuminances. There are three categories of road which are graded according to pedestrian and traffic use, and crime risk (Table 9.4).

There is a limited degree of glare control. This only applies to lanterns emitting more than 3500 lm in the lower hemisphere. In this case the maximum acceptable intensity at 80° is 160 cd/klm and at 90° is 80 cd/klm.

Table 9.4 Illuminance levels

Category	*Average Illum*	*Minimum Illum.*
1	10 lx	5 lx
2	6 lx	2.5 lx
3	2.5 lx	1 lx

Planning the layout

The steps are:

1. Decide on a type of layout (e.g. staggered), mounting height H, lantern type, and lamp.
2. Calculate the spacing S to achieve the minimum illuminance.

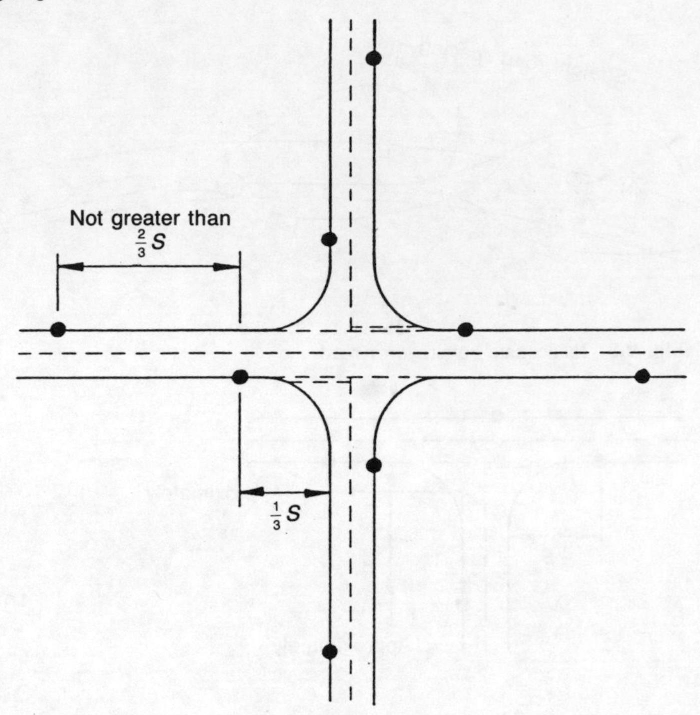

Fig. 9.10 Cross roads

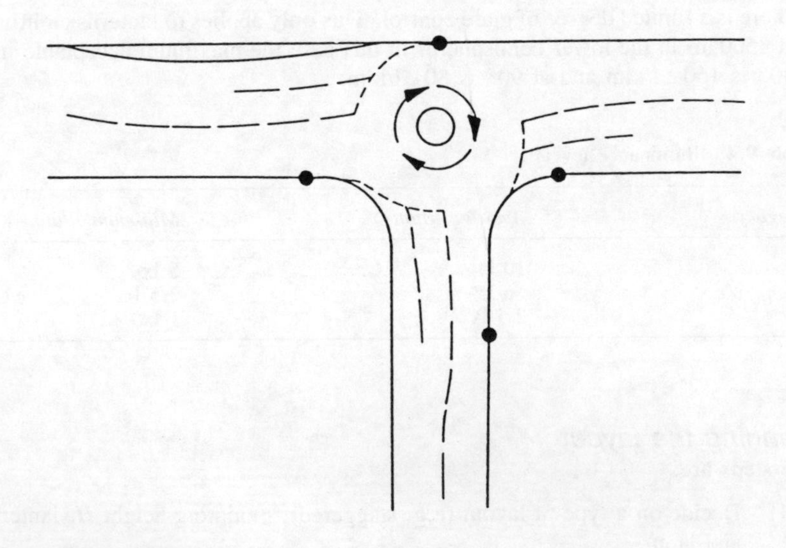

Fig. 9.11 Mini roundabout

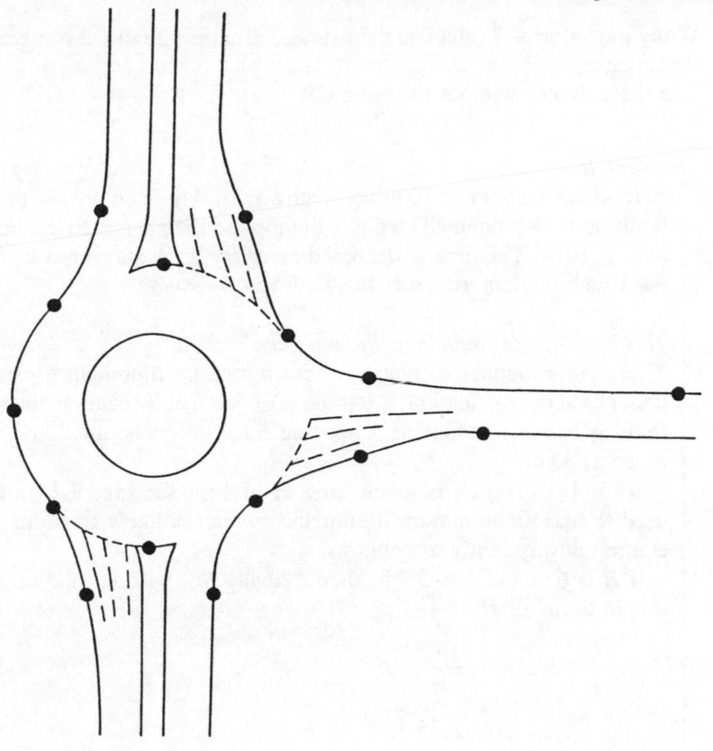

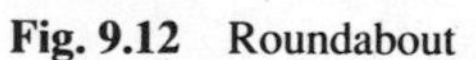

Fig. 9.12 Roundabout

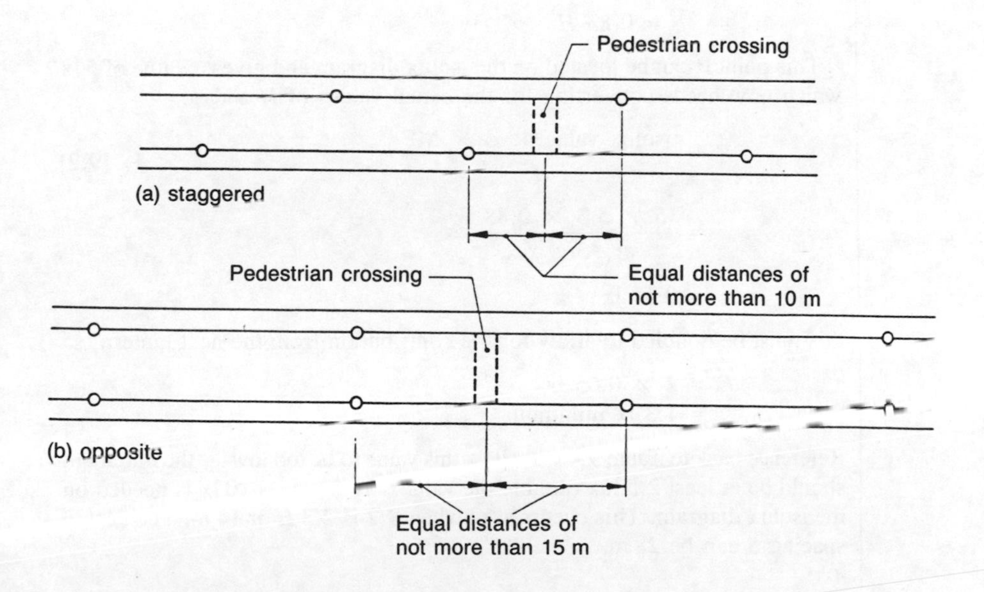

Fig. 9.13 Typical lantern positions adjacent to a pedestrian crossing

3. Using the value of S calculate the average illuminance to see if it complies with the category.
4. Site the columns without exceeding S.

Example

A road has a width of 10 m and pathway of 3 m. Lanterns with 70 W SON lamps are to be mounted on 6 m columns and the photometric data is as given in Fig. 9.14. The area is designed category 2. A staggered layout is to be used with the lanterns over the kerb (no overhang).

To calculate the minimum illuminance

There are a number of places where minimum illuminance can occur. To understand the calculation, it will be assumed that it occurs in the road centre, midway between columns. A spacing S has to be assumed and this will be taken as 33 m.

An isolux diagram is given for a H of 1 m (see Fig. 9.14). This can be used to find the minimum illuminance which is likely to occur in the road centre midway between columns.

If H is 6 m and S is 33 m, then midway $S/2$ will be 16.5 m. Expressing this in terms of H:

$$S/2 = \frac{16.5}{6} H$$

$$= 2.75\ H$$

The distance to the road centre will be:

$$W_1/2 = \frac{5}{6} H$$

$$= 0.83\ H$$

This point P can be located on the isolux diagram and gives a value of 5 lx which now needs converting for the actual values of H and ϕ:

$$E = \frac{\text{isolux value} \times \phi \times \text{MF}}{H^2} \quad \ldots [9.2]$$

$$= \frac{5 \times 5.3 \times 0.88}{36}$$

$$= 0.65 \text{ lx}$$

this must be doubled to allow for the contribution from the next lantern, so

$$E = 2 \times 0.65 \text{ lx}$$
$$= 1.3 \text{ lx minimum}$$

Reference back to Table 9.4 will show this value to be too low – the minimum should be at least 2.5 lux (double the value). This means 10 lx is needed on the isolux diagram. This occurs at Q when $S/2$ is 2.3 H or 14 m. The design spacing S can be 28 m.

To calculate the average illuminance (E)

Description: Road lighting lantern
Recommended Column Height(s): 5–6m

Version	Catalogue Number	Lamp
–	QB5 1050.4	50W SON-E
–	QB5 1070.4	70W SON-E

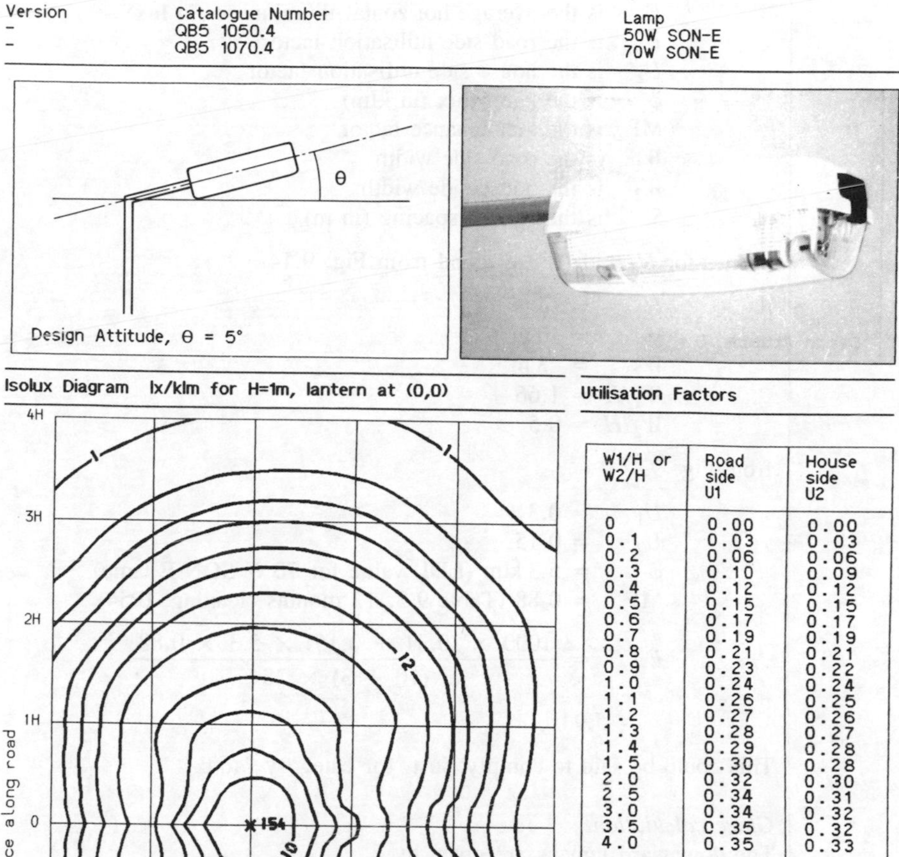

Isolux Diagram lx/klm for H=1m, lantern at (0,0)

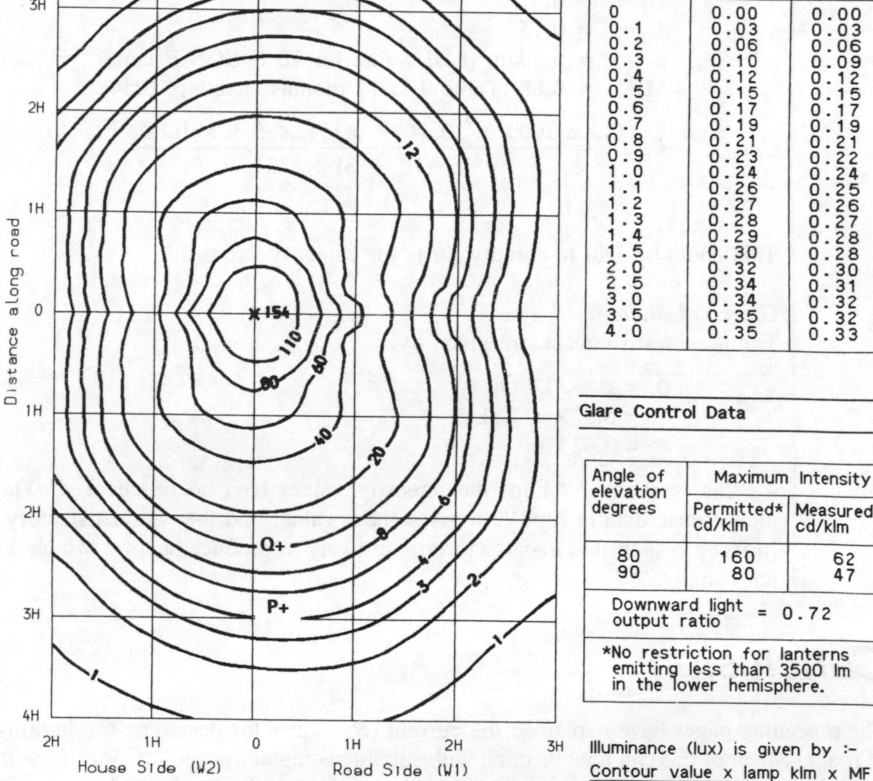

Utilisation Factors

W1/H or W2/H	Road side U1	House side U2
0	0.00	0.00
0.1	0.03	0.03
0.2	0.06	0.06
0.3	0.10	0.09
0.4	0.12	0.12
0.5	0.15	0.15
0.6	0.17	0.17
0.7	0.19	0.19
0.8	0.21	0.21
0.9	0.23	0.22
1.0	0.24	0.24
1.1	0.26	0.25
1.2	0.27	0.26
1.3	0.28	0.27
1.4	0.29	0.28
1.5	0.30	0.28
2.0	0.32	0.30
2.5	0.34	0.31
3.0	0.34	0.32
3.5	0.35	0.32
4.0	0.35	0.33

Glare Control Data

Angle of elevation degrees	Maximum Intensity	
	Permitted* cd/klm	Measured cd/klm
80	160	62
90	80	47
Downward light output ratio = 0.72		
*No restriction for lanterns emitting less than 3500 lm in the lower hemisphere.		

Illuminance (lux) is given by :–

$$\frac{\text{Contour value} \times \text{lamp klm} \times \text{MF}}{(\text{Mounting height H in m})^2}$$

Fig. 9.14 Photometric data for subsidiary road lantern (courtesy Thorn Lighting)

$$E = \frac{1000\ (U_1 + U_2) \times \phi \times \text{MF}}{(W_1 + W_2)\ \text{S}} \qquad \ldots [9.3]$$

where

E is the average horizontal illuminance (in lux)
U_1 is the road side utilisation factor
U_2 is the house side utilisation factor
ϕ is the lamp flux (in klm)
MF is the maintenance factor
W_1 is the road side width
W_2 is the house side width
S is the lantern spacing (in m)

Values for U_1 and U_2 are found from Fig. 9.14

$H = 6\,\text{m}$
$W_1 = 10\,\text{m}$
$W_2 = 3\,\text{m}$
$\therefore \quad W_1/H = 1.66$
$W_2/H = 0.5$

from Fig. 9.14

$U_1 = 0.31$
$U_2 = 0.15$
$\phi = 5.3$ klm (LDL value for 70 W SON-E lamp)
MF $= 0.88$ (Table 9.3, 12 months cleaning, IP54)

$$\therefore \quad E = \frac{1000 \times (0.31 + 0.15) \times 5.3 \times 0.88}{(10 + 3) \times 28}$$

$$= 5.9\,\text{lx}$$

This could be said to comply, just, for category 2 (6 lx)

Glare calculation
The downward lumens are given by:

$$\phi = \phi \times \text{DLOR}$$
$$= 5.3 \times 0.72\,\text{klm}$$
$$= 3.82\,\text{klm}$$

As this is above 3.6 klm, the intensity values have to be checked. The photometric data in Fig. 9.14 gives these values and they are satisfactory. In fact a responsible manufacturer is unlikely to produce lanterns which do not comply.

Conclusion

The preceding pages have explained the current techniques for designing the lighting of roads. Much of this can now be done with suitable computer programs. But, as with all lighting design, there is a range of options in choice of equipment, layout, and geometry. If this were not so, we need never remark 'well that's a well lit road'. They would all be the same!

10 Floodlighting

Floodlighting, as the term implies, is literally flooding a surface with light. If this is all that is needed then the design and calculations are based on achieving an acceptable level of illumination and uniformity on vertical surfaces, e.g. buildings, or on horizontal surfaces, e.g. car parks. This chapter will deal with these calculations.

There is often much more to outdoor lighting:

appearance during daytime
glare from the installation
decorative lighting
lighting for specific outdoor activities such as sport

A list of relevant publications which give guidance on these aspects is included at the end of the chapter.

Floodlighting a building

The general principles of illuminance calculations have been covered in earlier chapters. A practical example will now be taken to explain the design process.

Planning

Figure 10.1 shows a church. There is plenty of space available for siting the floodlights and even lighting is required on the whole elevation.

The basic requirements are the provision of even and sufficient illuminance. Successful floodlighting also requires a sense of drama and colour. In general, all floodlighting of a flat plain surface should be reasonably uniform. There is a greater degree of tolerance than for an interior lighting scheme, but all the beams from each floodlight should overlap. A suggested uniformity ratio max : av is 5 : 1.

Draw a series of overlapping circles to cover the whole elevation with the minimum of waste light. This is shown in Fig. 10.2.

A circle represents a beam of light whose edge is about 1/10th the illuminance of the centre, these circles should be drawn so that the edge of one passes through the centre of the next beam. Provided that the floodlights are a reasonable distance away and have symmetrical beams this technique will produce an even floodlighting effect.

Fig. 10.1

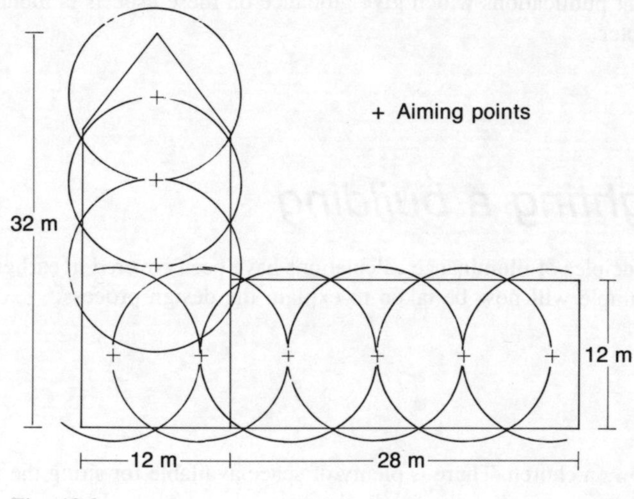

Fig. 10.2

In this example, nine floodlights will provide coverage of the church and tower. If the distance from the floodlights to the church is 50 m, and as each beam will be about 14 m wide, the beam width will be given by:

$$\tan \frac{\text{(beam angle)}}{2} = \frac{\text{radius of beam}}{\text{distance}} \qquad \ldots [10.1]$$

and in this case:

$$= \frac{7}{50}$$

$$\therefore \text{ beam angle} = 2 \tan^{-1}(7/50) = 16°$$

Floodlights can be categorised by their beam angles, e.g. narrow beam, but these classifications are sometimes vague and it is better to work with actual beam angles, as these are normally quoted by the supplier.

A typical set of floodlighting data is given in Table 10.1. This is from Philips Catalogue, NNF 020 Floodlight range, and has some narrow beam types. NNF 020/1 has a 14° beam (2 × 7°) and 250 W SON/T lamp gives 27 000 lm and looks a possible solution.

Table 10.1 Beam data (courtesy Philips Lighting Ltd)

Floodlight	*NNF 020/1 400W SON Narrow Beam*	*NNF 020/2 400W SON Wide Beam*	*NNF 020/1 250W SON/T Narrow Beam*	*NNF 020/2 250W SON/T Wide Beam*
Peak Intensity cd/1000 lm	1455	1416	20500	7585
Beam Efficiency Total Beam	68%	76%	60%	62%
Beam Angles (10% peak)				
Vertical	2 × 37°	2 × 30°	2 × 7°	2 × 14°
Horizontal	2 × 37°	2 × 39°	2 × 7°	2 × 14°
Beam Angles (50% peak)				
Vertical	2 × 15°	2 × 17°	2 × 2°	2 × 4°
Horizontal	2 × 15°	2 × 17°	2 × 2°	2 × 4°

At this stage it is worth checking the illuminance at the beam peak (the maximum). It is 20 500 cd/1000 lm.

$$\therefore \quad \text{Actual intensity} = 20\,000 \times \frac{27\,000}{1000}$$

$$= 553\,500 \text{ cd}$$

$$\therefore \quad \text{Peak (or max) illuminance} = \frac{553\,500}{(\text{dist})^2}$$

$$= \frac{553\,500}{50^2}$$

$$= 220 \text{ lx}$$

Applying a light loss factor of 0.7

$$\text{Peak illuminance} = 220 \times 0.7 = 154 \text{ lx}$$

Illuminance levels

The amount of light will depend on a number of factors. The basic requirement is that the floodlit area will stand out from its surroundings. Table 10.2 gives a guide on illuminance values, assuming a building reflectance of 0.35 (medium stone).

Table 10.2 Recommended Building Illuminances District Brightness (courtesy Philips Lighting Ltd)

Building Condition	*High*	*Medium*	*Low*
Clean	90	55	35
Fairly Clean	140	90	50
Dirty	280	180	100

If, in this example, the church is on the outskirts of a town and is made of stone in a fair state of cleanliness, the peak illuminance of 154 lx is a reasonable design level.

Lumen calculation

To check the average illuminance a basic lumen calculation is used.

$$\text{Average illuminance} = \frac{N \times L \times \text{LLF} \times \text{BF} \times \text{WLF}}{\text{total surface area}} \quad \ldots [10.2]$$

where N = number of floodlights
L = initial lamp lumens per floodlight
LLF = light loss factor
BF = beam lumens/lamp lumens (beam efficiency)
WLF = waste light factor – some of the beam will miss the building and this figure is taken as about 0.7 of the light reaching its target

$$\therefore \quad E = \frac{9 \times 27\,000 \times 0.7 \times 0.6 \times 0.7}{(12 \times 32) + (12 \times 28)}$$

$$= 100\,\text{lx}$$

Again this value would appear to be satisfactory. As already stated, this is just part of designing a floodlighting scheme. The following is a list of other considerations:

Modelling – the three dimensional shape should be revealed by suitable direction of the floodlights.
Colour – the colour of the lamp, if feasible, should be different to those used in the general environment.
Glare – beams must be concealed from the principal directions of view.

Floodlighting a horizontal open area

The process for lighting a horizontal area is similar to that used for the church, but it is more accurate to use horizontal illuminance plots such as isolux diagrams (see Chapter

2). In the next example the requirement is to illuminate a large car park using floodlights on 8 m high poles. An illuminance of 20 lux is required (CIBSE recommend 5–20 lx).

Planning

Figure 10.3 gives the isolux data for a 400 W MBF floodlight on a flat plane 8 m below the floodlight. This is the distribution of illuminance on a plane parallel to the front of the floodlight and normal to the axis of the beam. The dotted line on the figure shows the boundary where the illuminance has fallen to 1/10th of the maximum, i.e. this is the approximate area that a single floodlight would cover. The average illuminance would not be the maximum but would lie somewhere between the maximum and boundary value.

It is important to appreciate the isolux diagrams refer to a specific distance and direction. They can, however, be used for other distances and planes at different angles. This involves the use of the inverse square law and cosine law. If the distance is increased, then the illuminance will decrease but the areas will increase correspondingly.

A simplified method of design can be set out as follows:

(a) Choose the most suitable type of lamp and wattage for the application.
(b) Choose the most suitable type of floodlight.

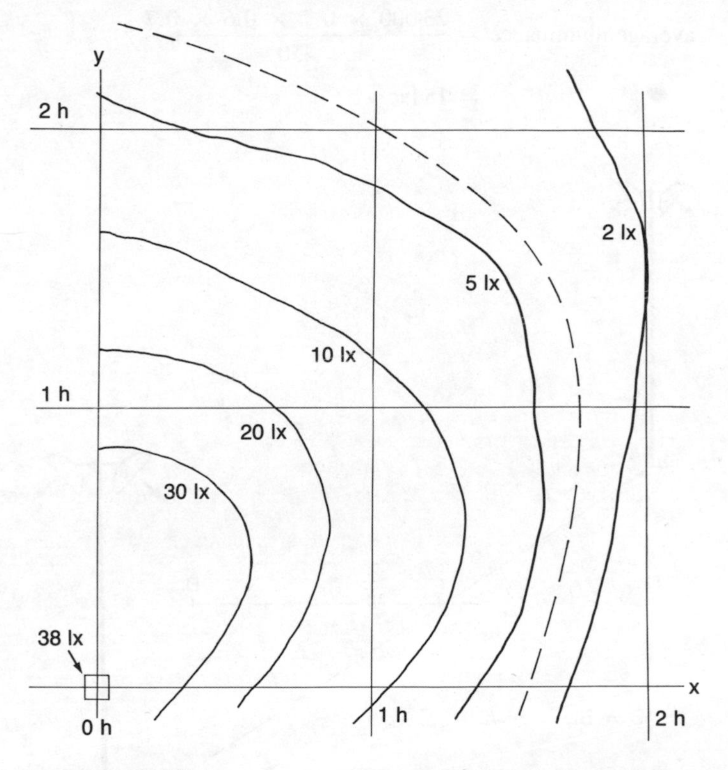

Fig. 10.3

(c) Choose the mounting height to be used.
(d) Determine the maximum spacing to achieve an acceptable diversity.
(e) Check that the scheme will provide the design level of average illuminance.

In this case the choice is between a cluster of floodlights at the top of each pole, or a single post top lantern (street lighting lanterns are not satisfactory as their light distribution is not designed for this situation). High wattage lamps can be used and the initial design is two 400 W MBF floodlights mounted back to back. Their isolux diagram for 8 m mounting is shown in Fig. 10.3 and a diversity in illuminance of 10 : 1 should be acceptable.

Using a layout as illustrated in Fig. 10.4 the illuminance directly below each pair of floodlights will be 76 lux. At point B midway between two poles the illuminance needs to be at least 7.6 lux for a diversity of 10 : 1. To achieve this, each flood must provide 7.6/4 lux which will occur when AB is approximately 15 m. Hence the spacing AC will be 30 m.

Next consider the distance along the car park. With 30 m spacing, check the illuminance at the mid-point D. This will be 2 lx from each floodlight giving a total of 8 lx. This is satisfactory (it exceeds the 7.6 lx requirement).

The average illuminance can be checked using the lumen method. Each floodlight effectively lights an area 30 m wide × 15 m across, i.e. 450 sq m. Using the formula on p. 180 and assuming the same values for LLF, BF, and WLF:

$$\text{average illuminance} = \frac{23\,000 \times 0.7 \times 0.6 \times 0.7}{450}$$

$$= 15\ \text{lx}$$

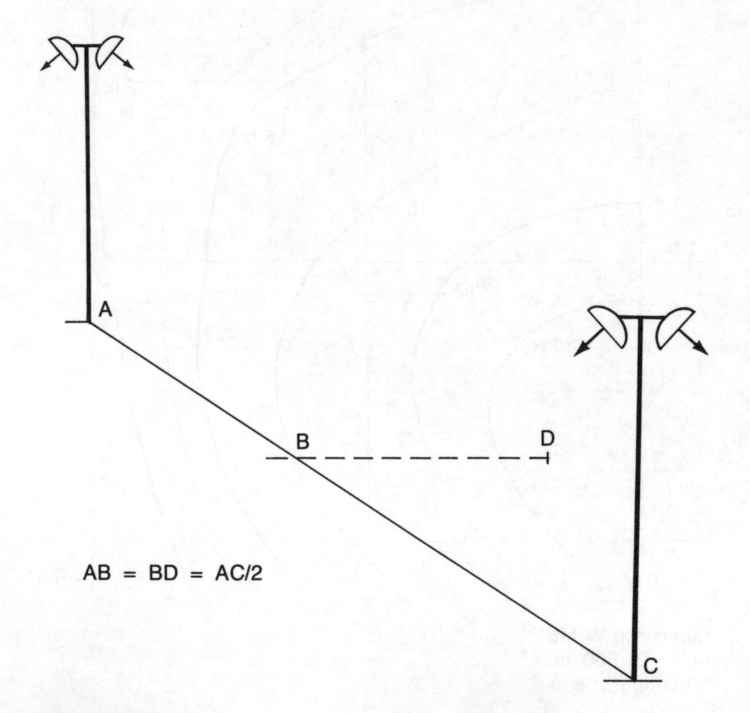

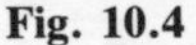
Fig. 10.4

This is below the design requirements of 20 lx. In practice, due to the diversity of 10 : 1 it is doubtful if the design would be altered. It is unrealistic to be too precise when assumptions, like the factors, are made.

To get a 20 lx *average* illuminance the poles would have to be spaced closer or, if this is unsatisfactory, then another floodlight would have to be considered. The area to be covered by each floodlight would be reduced to 450 × 15/20 or 337 m^2. Another solution could be to use 360 W SON-E lamps, if suitable. This would raise the illuminance by 40 000/23 000 to 26 lx. This assumes that the floodlight will take either type of lamp.

Unfortunately suppliers of floodlighting equipment may not provide isolux diagrams in their technical data. The choice then is either to draw a diagram from the intensity distribution data, or ask the photometric department of the supplier for the diagram.

Sources of specialised data

Building floodlighting

Electricity Council	– Outdoor Lighting	1983
CIBSE	– The Outdoor Environment	1975

Outdoor Areas

CIBSE Guides	– Building & Civil Engineering Sites	1975
	Shipbuilding & Ship Repair	1979
	The Industrial Environment	1989

Outdoor Sports

CIBSE Guide	– Sports	1990

Appendix: Range of typical lamps and their lumen outputs

Lamp type	Wattage	Average initial lumens	Lighting design lumens
Tungsten filament		(100 h)	(2000 h)
GLS	60	610	–
	150	2 060	–
	500	8 200	–
Tungsten			
Halogen K	750	15 000	–
	1 500	32 000	–
Fluorescent tubes			
White	75	6 050	5 750
	65	5 100	4 750
	40	3 050	2 800
Natural	75	4 400	4 000
	65	3 700	3 400
	40	2 300	2 100
Krypton filled 26mm			
Triphosphor	70	6 550	6 300
	58	5 400	5 100
	36	3 450	3 200
Compact lamps	16	1 050	1 800
	24	1 800	1 690
Low-pressure sodium			
SOX	35	4 600	4 500
	180	32 000	31 500
SLI	60	6 000	5 700
	200	20 500	20 000
High-pressure sodium			
SON	70	6 000	5 500
	400	47 000	45 000
SON Deluxe	150	12 500	11 500
	400	38 500	35 500
High-pressure mercury			
MBF	50	1 900	1 750
	125	6 200	5 700
	1 000	58 000	54 000
MBIF	250	19 000	16 000
	1 000	92 000	85 000

Bibliography

A comparatively full list of references is contained in the CIBS Code 1984. The following are relevant books, CIBS publications, recommendations, and regulations that may be of help to the student.

Books

Lamps and Lighting, 3rd edn. M A Cayless and A M Marsden (eds). Arnold 1983
Sight, Light and Work, H C Weston. Lewis 1962
Interior Lighting Design Handbook. Electricity Council
Day Lighting, R G Hopkinson, P Petherbridge, J Longmore. Heinemann 1963
Lighting Fittings – Design & Performance, A R Bean and R H Simons. Pergamon 1968
Principles of Natural Lighting, J A Lynes. Elsevier 1968
Lighting Design in Buildings, J Boud. Peter Peregrinus 1973

HMSO Publications

Architectural Physics: Lighting. R G Hopkinson
BRE, MPBW: *BRE Daylight Protractors*

British Standards and Codes of Practice

BS CP3	Chapter 1: Day lighting. Part 1: Day lighting
BS CP3	Chapter 1: Lighting. Part 2: Artificial lighting
BS CP3	Chapter 1 (B): Sunlight (houses, flats and schools)
BS CP3	Chapter VII (F): Provision of artificial light (houses, flats and schools)
BS 889	*Flameproof electric lighting fittings*
BS 1853	Parts 1 and 2: *Tubular fluorescent lamps for general lighting service*
BS 2560	*Exit signs (internally illuminated)*
BS 4533	Parts 1, 2, 101 and 102: *Luminaires*
BS 4683	Parts 1–4: *Electrical apparatus for explosive atmospheres*
BS 4727	Part 4: *Glossary of terms particular to lighting and colour*
BS 4800	*Paint colours for building purposes*
BS 4901	*Plastics colours for building purposes*
BS 5225	Part 1: *Photometric measurements for luminaires*
BS 5252	*Framework for colour coordination for building purposes*
BS 5266	Parts 1 and 3: *Emergency lighting*

BS 5345 Parts 1–8: *Code of practice for the selection installation and maintenance of electrical apparatus for use in potentially explosive atmospheres (other than mining applications or explosive processing and manufacture)*
BS 5489 *Code of practice for road lighting*
BS 5490 *Classification of degrees of protection provided by enclosures*
BS DD67 *Basic data for the design of buildings: sunlight*
BS DD73 *Basic data for the design of buildings: daylight*

Statutory rules

Factories Act 1961
Offices, Shops and Railway Premises Act 1963
Building Regulations 1965
Fire Precautions Acts 1971

CIBSE publications

Codes for Interior Lighting, 1984 (commonly known as CIBSE 1984 Lighting Code)

Technical memoranda

TM5 *The calculation and use of utilisation factors*
TM10 *The calculation of glare indices*
TM12 *Emergency Lighting*

Lighting guides

Hostile and hazardous environments
Hospitals and health care buildings
Museums and art galleries
Shipbuildings and ship repair
Window design
The industrial environment
Areas for visual display terminals

Lighting Industry Federation Factfinders

No. 2 *Dimming*
No. 3 *Lamp guide*
No. 4 *Energy and lighting*
No. 5 *The benefits of certification*
No. 6 *Hazardous area lighting*

Recommendations and reports of the Commission Internationale de l'Eclairage

No. 13.2 *Method of measuring and specifying colour rendering of light sources* 1974
No. 17 *International lighting vocabulary* 1970
No. 18 *Principles of light measurements* 1970
No. 19/2 *An analytic model for describing the influence of lighting parameters on visual performance* 1981
No. 24 *Photometry of indoor type luminaires with tubular fluorescent lamps* 1973
No. 25 *Procedures for the measurement of luminous flux of discharge lamps and for their calibration as working standards* 1973
No. 29 *Guide on interior lighting* 1975
No. 40 *Calculations for interior lighting – basic method* 1978
No. 41 *Light as a true visual quantity: principles of measurement* 1978
No. 46 *A review of publications on properties and reflection values of material reflection standards* 1979

Index